Intermediate Probability Theory for Biomedical Engineers

Intermediate Probability Theory for Biomedical Engineers
John D. Enderle, David C. Farden, and Daniel J. Krause

ISBN: 978-3-031-00486-5   paperback
ISBN: 978-3-031-01614-1   ebook

DOI 10.1007/978-3-031-01614-1

A Publication in the Springer series
SYNTHESIS LECTURES ON BIOMEDICAL ENGINEERING #10

Lecture #10
Series Editor: John D. Enderle, University of Connecticut

Series ISSN: 1930-0328   print

Series ISSN: 1930-0336   electronic

First Edition
10 9 8 7 6 5 4 3 2 1

Intermediate Probability Theory for Biomedical Engineers

John D. Enderle
Program Director & Professor for Biomedical Engineering
University of Connecticut

David C. Farden
Professor of Electrical and Computer Engineering
North Dakota State University

Daniel J. Krause
Emeritus Professor of Electrical and Computer Engineering
North Dakota State University

SYNTHESIS LECTURES ON BIOMEDICAL ENGINEERING #10

ABSTRACT

This is the second in a series of three short books on probability theory and random processes for biomedical engineers. This volume focuses on *expectation, standard deviation, moments*, and the *characteristic function*. In addition, *conditional expectation, conditional moments* and the *conditional characteristic function* are also discussed. *Jointly distributed random variables* are described, along with *joint expectation, joint moments*, and the *joint characteristic function. Convolution* is also developed. A considerable effort has been made to develop the theory in a logical manner—developing special mathematical skills as needed. The mathematical background required of the reader is basic knowledge of differential calculus. Every effort has been made to be consistent with commonly used notation and terminology—both within the engineering community as well as the probability and statistics literature. The aim is to prepare students for the application of this theory to a wide variety of problems, as well give practicing engineers and researchers a tool to pursue these topics at a more advanced level. Pertinent biomedical engineering examples are used throughout the text.

KEYWORDS

Probability Theory, Random Processes, Engineering Statistics, Probability and Statistics for Biomedical Engineers, Statistics. Biostatistics, Expectation, Standard Deviation, Moments, Characteristic Function

Contents

Preface

This is the second in a series of short books on probability theory and random processes for biomedical engineers. This text is written as an introduction to probability theory. The goal was to prepare students at the sophomore, junior or senior level for the application of this theory to a wide variety of problems—as well as pursue these topics at a more advanced level. Our approach is to present a unified treatment of the subject. There are only a few key concepts involved in the basic theory of probability theory. These key concepts are all presented in the first chapter. The second chapter introduces the topic of random variables. Later chapters simply expand upon these key ideas and extend the range of application.

This short book focuses on expectation, standard deviation, moments, and the characteristic function. In addition, conditional expectation, conditional moments and the conditional characteristic function are also discussed. Jointly distributed random variables are described, along with joint expectation, joint moments, and the joint characteristic function. Convolution is also developed.

A considerable effort has been made to develop the theory in a logical manner—developing special mathematical skills as needed. The mathematical background required of the reader is basic knowledge of differential calculus. Every effort has been made to be consistent with commonly used notation and terminology—both within the engineering community as well as the probability and statistics literature.

The applications and examples given reflect the authors' background in teaching probability theory and random processes for many years. We have found it best to introduce this material using simple examples such as dice and cards, rather than more complex biological and biomedical phenomena. However, we do introduce some pertinent biomedical engineering examples throughout the text.

Students in other fields should also find the approach useful. Drill problems, straightforward exercises designed to reinforce concepts and develop problem solution skills, follow most sections. The answers to the drill problems follow the problem statement in random order. At the end of each chapter is a wide selection of problems, ranging from simple to difficult, presented in the same general order as covered in the textbook.

We acknowledge and thank William Pruehsner for the technical illustrations. Many of the examples and end of chapter problems are based on examples from the textbook by Drake [9].

CHAPTER 3

Expectation

Suppose that an experiment is performed N times and the RV x is observed to take on the value $x = x_i$ on the ith trial, $i = 1, 2, \ldots, N$. The average of these N numbers is

$$\overline{x}_N = \frac{1}{N} \sum_{i=1}^{N} x_i. \tag{3.1}$$

We anticipate that as $N \to \infty$, the average observed value of the RV x would converge to a constant, say $\overline{x}$. It is important to note that such sums do not always converge; here, we simply appeal to one's intuition to suspect that convergence occurs. Further, we have the intuition that the value $\overline{x}$ can be computed if the CDF F_x is known. For example, if a single die is tossed a large number of times, we expect that the average value on the face of the die would approach

$$\frac{1}{6}(1 + 2 + 3 + 4 + 5 + 6) = 3.5.$$

For this case we predict

$$\overline{x} = \sum_{i=1}^{6} i P(x = i) = \int_{-\infty}^{\infty} \alpha \, dF_x(\alpha). \tag{3.2}$$

A little reflection reveals that this computation makes sense even for continuous RVs: the predicted value for $\overline{x}$ should be the "sum" of all possible values the RV x takes on weighted by the "relative frequency" or probability the RV takes on that value. Similarly, we predict that the average observed value of a function of x, say $g(x)$, to be

$$\overline{g(x)} = \int_{-\infty}^{\infty} g(\alpha) \, dF_x(\alpha). \tag{3.3}$$

Of course, whether or not this prediction is realized when the experiment is performed a large number of times depends on how well our model for the experiment (which is based on probability theory) matches the physical experiment.

The statistical average operation performed to obtain $g(\overline{x})$ is called **statistical expectation**. The sample average used to estimate $\overline{x}$ with $\overline{x}_N$ is called the **sample mean**. The quality of estimate attained by a sample mean operation is investigated in a later chapter. In this chapter, we present definitions and properties of statistical expectation operations and investigate how knowledge of certain moments of a RV provides useful information about the CDF.

3.1   MOMENTS

Definition 3.1.1.   *The **expected value** of $g(x)$ is defined by*

$$E(g(x)) = \int_{-\infty}^{\infty} g(\alpha)\, dF_x(\alpha)\,, \tag{3.4}$$

*provided the integral exists. The **mean of the RV** x is defined by*

$$\eta_x = E(x) = \int_{-\infty}^{\infty} \alpha\, dF_x(\alpha). \tag{3.5}$$

*The **variance of the RV** x is defined by*

$$\sigma_x^2 = E((x - \eta_x)^2), \tag{3.6}$$

*and the nonnegative quantity σ_x is called the **standard deviation**. The **nth moment** and the **nth central moment**, respectively, are defined by*

$$m_n = E(x^n) \tag{3.7}$$

and

$$\mu_n = E((x - \eta_x)^n). \tag{3.8}$$

The expected value of $g(x)$ provides some information concerning the CDF F_x. Knowledge of $E(g(x))$ does not, in general, enable F_x to be determined—but there are exceptions. For any real value of α,

$$E(u(\alpha - x)) = \int_{-\infty}^{\infty} u(\alpha - \alpha')\, dF_x(\alpha') = \int_{-\infty}^{\alpha} dF_x(\alpha') = F_x(\alpha). \tag{3.9}$$

The sample mean estimate for $E(u(\alpha - x))$ is

$$\frac{1}{n} \sum_{i=1}^{n} u(\alpha - x_i),$$

the empirical distribution function discussed in Chapter 2. If $\alpha \in \Re^*$ and x is a continuous RV then (for all α where f_x is continuous)

$$E(\delta(\alpha - x)) = \int_{-\infty}^{\infty} \delta(\alpha - \alpha') f_x(\alpha') \, d\alpha' = f_x(\alpha). \qquad (3.10)$$

Let A be an event on the probability space (S, F, P), and let

$$I_A(\zeta) = \begin{cases} 1, & \text{if } \zeta \in A \\ 0, & \text{otherwise.} \end{cases} \qquad (3.11)$$

With $x(\zeta) = I_A(\zeta)$, x is a legitimate RV with $x^{-1}(\{1\}) = A$ and $x^{-1}(\{0\}) = A^c$. Then

$$E(x) = \int_{-\infty}^{\infty} \alpha \, dF_x(\alpha) = P(A). \qquad (3.12)$$

The above result may also be written in terms of the Lebesgue-Stieltjes integral as

$$E(I_A(\zeta)) = \int_{\zeta \in S} I_A(\zeta) dP(\zeta) = \int_A dP(\zeta) = P(A). \qquad (3.13)$$

The function I_A is often called an **indicator function**.

If one interprets a PDF f_x as a "mass density", then the mean $E(x)$ has the interpretation of the center of gravity, $E(x^2)$ becomes the moment of inertia about the origin, and the variance σ_x^2 becomes the central moment of inertia. The standard deviation σ_x becomes the radius of gyration. A small value of σ_x^2 indicates that most of the mass (probability) is concentrated at the mean; i.e., $x(\zeta) \approx \eta_x$ with high probability.

Example 3.1.1.   *The RV x has the PMF*

$$p_x(\alpha) = \begin{cases} \frac{1}{4}, & \alpha = b - a \\ \frac{1}{4}, & \alpha = b + a \\ \frac{1}{2}, & \alpha = b \\ 0, & \text{otherwise,} \end{cases}$$

where a and b are real constants with $a > 0$. Find the mean and variance for x.

Solution.   We obtain

$$\eta_x = E(x) = \int_{-\infty}^{\infty} \alpha \, dF_x(\alpha) = \frac{b-a}{4} + \frac{b}{2} + \frac{b+a}{4} = b$$

and

$$\sigma_x^2 = E((x - \eta_x)^2) = \int_{-\infty}^{\infty} (\alpha - \eta_x)^2 \, dF_x(\alpha) = \frac{a^2}{2}.$$

∎

Example 3.1.2. *The RV x has PDF*

$$f_x(\alpha) = \frac{1}{b - a}(u(\alpha - a) - u(\alpha - b))$$

where a and b are real constants with a < b. Find the mean and variance for x.

Solution. We have

$$E(x) = \frac{1}{b - a} \int_a^b \alpha \, d\alpha = \frac{b^2 - a^2}{2(b - a)} = \frac{b + a}{2}$$

and

$$\sigma_x^2 = \frac{1}{b - a} \int_a^b \left(\alpha - \frac{b + a}{2} \right)^2 d\alpha = \frac{1}{b - a} \int_{-(b-a)/2}^{(b-a)/2} \beta^2 \, d\beta = \frac{(b - a)^2}{12}.$$

∎

Example 3.1.3. *Find the expected value of $g(x) = 2x^2 - 1$, where*

$$f_x(\alpha) = \begin{cases} \dfrac{1}{3}\alpha^2, & -1 < \alpha < 2 \\ 0, & \text{otherwise.} \end{cases}$$

Solution. By definition,

$$E(g(x)) = \int_{-\infty}^{+\infty} g(\alpha) f_x(\alpha) \, d\alpha = \frac{1}{3} \int_{-1}^{2} (2\alpha^2 - 1)\alpha^2 \, d\alpha = \frac{17}{5}.$$

∎

Example 3.1.4. *The RV x has PDF*

$$f_x(\alpha) = \begin{cases} 1.5(1 - \alpha^2), & 0 \le \alpha < 1 \\ 0, & \text{elsewhere.} \end{cases}$$

Find the mean, the second moment, and the variance for the RV x.

Solution. From the definition of expectation,

$$\eta_x = E(x) = \int_{-\infty}^{\infty} \alpha f_x(\alpha) \, d\alpha = \frac{3}{2} \int_0^1 (\alpha - \alpha^3) \, d\alpha = \frac{3}{8}.$$

Similarly, the second moment is

$$m_2 = E(x^2) = \int_{-\infty}^{\infty} \alpha^2 f_x(\alpha)\, d\alpha = \frac{3}{2} \int_0^1 (\alpha^2 - \alpha^4)\, d\alpha = \frac{3}{2}\left(\frac{5-3}{15}\right) = \frac{1}{5}.$$

Applying the definition of variance,

$$\sigma_x^2 = \int_0^1 \left(\alpha - \frac{3}{8}\right)^2 \frac{3}{2}(1 - \alpha^2)\, d\alpha.$$

Instead of expanding the integrand directly, it is somewhat easier to use the change of variable $\beta = \alpha - \frac{3}{8}$, to obtain

$$\sigma_x^2 = \frac{3}{2} \int_{-3/8}^{5/8} \left(\frac{55}{64}\beta^2 - \frac{3}{4}\beta^3 - \beta^4\right) d\beta = 0.059375.$$

The following theorem and its corollary provide an easier technique for finding the variance. The result of importance here is

$$\sigma_x^2 = E(x^2) - \eta_x^2 = \frac{1}{5} - \left(\frac{3}{8}\right)^2 = \frac{19}{320} = 0.059375.$$

The PDF for this example is illustrated in Fig. 3.1. Interpreting the PDF as a mass density along the abscissa, the mean is the center of gravity. Note that the mean always falls between the minimum and maximum values for which the PDF is nonzero. ∎

The following theorem establishes that expectation is a linear operation and that the expected value of a constant is the constant.

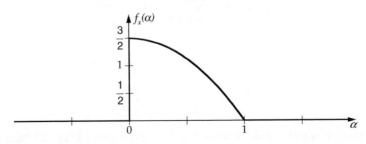

FIGURE 3.1: PDF for Example 3.1.4.

Theorem 3.1.1. *The expectation operator satisfies*

$$E(a) = a \tag{3.14}$$

and

$$E(a_1 g_1(x) + a_2 g_2(x)) = a_1 E(g_1(x)) + a_2 E(g_2(x)), \tag{3.15}$$

where a, a_1, and a_2 are arbitrary constants and we have assumed that all indicated integrals exist.

Proof. The desired results follow immediately from the properties of the Riemann-Stieltjes integral and the definition of expectation. ∎

Applying the above theorem, we find

$$\sigma_x^2 = E((x - \eta_x)^2) = E(x^2 - 2\eta_x x + \eta_x^2) = E(x^2) - \eta_x^2, \tag{3.16}$$

as promised in Example 3.1.4. The following corollary provides a general relationship between moments and central moments.

Corollary 3.1.1. *The nth central moment for the RV x can be found from the moments $\{m_0, m_1, \ldots, m_n\}$ as*

$$\mu_n = E((x - \eta_x)^n) = \sum_{k=0}^{n} \binom{n}{k} m_k (-\eta_x)^{n-k}. \tag{3.17}$$

Similarly, the nth moment for the RV x can be found from the central moments $\{\mu_0, \mu_1, \ldots, \mu_n\}$ as

$$m_n = E(x^n) = \sum_{k=0}^{n} \binom{n}{k} \mu_k (\eta_x)^{n-k}. \tag{3.18}$$

Proof. From the Binomial Theorem, we have for any real constant a:

$$(x - a)^n = \sum_{k=0}^{n} \binom{n}{k} x^k (-a)^{n-k}$$

and

$$x^n = ((x - a) + a)^n = \sum_{k=0}^{n} \binom{n}{k} (x - a)^k a^{n-k}.$$

Taking the expected value of both sides of the above equations and using the fact that expectation is a linear operation, the desired results follow by choosing $a = \eta_x$. ∎

In many advanced treatments of probability theory (e.g. [4, 5, 11]), expectation is defined in terms of the Lebesgue-Stieltjes integral

$$E(g(x)) = \int_S g(x(\zeta))\, dP(\zeta). \tag{3.19}$$

In most cases (whenever the Lebesgue-Stieltjes integral and the Riemann-Stieltjes integral both exist) the two definitions yield identical results. The existence of the Lebesgue-Stieltjes integral (3.19) requires

$$E(|g(x)|) = \int_S |g(x(\zeta))|\, dP(\zeta) < \infty, \tag{3.20}$$

whereas the Riemann-Stieltjes integral (3.4) may exist even though

$$E(|g(x)|) = \int_{-\infty}^{\infty} |g(\alpha)|\, dF_x(\alpha) = \infty. \tag{3.21}$$

Consequently, using (3.4) as a definition, we will on occasion arrive at a value for $E(g(x))$ in cases where $E(|g(x)|) = \infty$. There are applications for which this more liberal interpretation is useful.

Example 3.1.5. *Find the mean and variance of the RV x with PDF*

$$f_x(\alpha) = \frac{1}{\pi(1 + \alpha^2)}.$$

Solution. By definition,

$$\eta_x = \lim_{T_1, T_2 \to \infty} \int_{-T_1}^{T_2} \alpha f_x(\alpha)\, d\alpha\,,$$

assuming the limit exists independent of the manner in which $T_1 \to \infty$ and $T_2 \to \infty$. For this example, we have

$$\int_{-T_1}^{T_2} \alpha f_x(\alpha)\, d\alpha = \frac{1}{2\pi}(\ln(1 + T_2^2) - \ln(1 + T_1^2)).$$

Consequently, the limit indicated above does not exist. If we restrict the limit to the form $T_1 = T_2 = T$ (corresponding to the Cauchy principle value of the integral) then we obtain

$\eta_x = 0$. Accepting $\eta_x = 0$ for the mean, we find

$$E(x^2) = \lim_{T_1, T_2 \to \infty} \int_{-T_1}^{T_2} \alpha^2 f_x(\alpha)\, d\alpha = +\infty,$$

and we conclude that $\sigma_x^2 = \infty$.   ∎

The computation of high order moments using the direct application of the definition (3.4) is often tedious. We now explore some alternatives.

Example 3.1.6.   *The RV x has PDF $f_x(\alpha) = e^{-\alpha} u(\alpha)$. Express m_n in terms of m_{n-1} for $n = 1, 2, \ldots$.*

Solution.   By definition, we have

$$m_n = E(x^n) = \int_0^\infty \alpha^n e^{-\alpha}\, d\alpha.$$

Integrating by parts (with $u = \alpha^n$ and $dv = e^{-\alpha}\, d\alpha$)

$$m_n = -\alpha^n e^{-\alpha}\Big|_0^\infty + n \int_0^\infty \alpha^{n-1} e^{-\alpha}\, d\alpha = n m_{n-1}, \quad n = 1, 2, \ldots.$$

Note that $m_0 = E(1) = 1$. For example, we have $m_4 = 4 \cdot 3 \cdot 2 \cdot 1 = 4!$. We have used the fact that for $n > 0$

$$\lim_{\alpha \to \infty} \alpha^n e^{-\alpha} = 0.$$

This can be shown by using the Taylor series for e^α to obtain

$$\frac{\alpha^n}{e^\alpha} = \frac{\alpha^n}{\displaystyle\sum_{k=0}^\infty \frac{\alpha^k}{k!}} \le \frac{\alpha^n}{\dfrac{\alpha^{n+1}}{(n+1)!}} = \frac{(n+1)!}{\alpha}$$ ∎

The above example illustrates one technique for avoiding tedious repeated integration by parts. The moment generating function provides another frequently useful escape, trading repeated integration by parts with repeated differentiation.

Definition 3.1.2.   *The function*

$$M_x(\lambda) = E(e^{\lambda x}) \tag{3.22}$$

*is called the **moment generating function for the RV** x, where λ is a real variable.*

Although the moment generating function does not always exist, when it does exist, it is useful for computing moments for a RV, as shown below. In Section 3.3 we introduce a related function, the characteristic function. The characteristic function always exists and can also be used to obtain moments.

Theorem 3.1.2. *Let $M_x(\lambda)$ be the moment generating function for the RV x, and assume $M_x^{(n)}(0)$ exists, where*

$$M_x^{(n)}(\lambda) = \frac{d^n M_x(\lambda)}{d\lambda^n}. \tag{3.23}$$

Then

$$E(x^n) = M_x^{(n)}(0). \tag{3.24}$$

Proof. Noting that

$$\frac{d^n e^{\lambda x}}{d\lambda^n} = x^n e^{\lambda x}$$

we have $M_x^{(n)}(\lambda) = E(x^n e^{\lambda x})$. The desired result follows by evaluating at $\lambda = 0$. ∎

Example 3.1.7. *The RV x has PDF $f_x(\alpha) = e^{-\alpha} u(\alpha)$. Find $M_x(\lambda)$ and $E(x^n)$, where n is a positive integer.*

Solution. We find

$$M_x(\lambda) = \int_0^\infty e^{(\lambda-1)\alpha}\, d\alpha = \frac{1}{1-\lambda},$$

provided that $\lambda < 1$. Straightforward computation reveals that

$$M_x^{(n)}(\lambda) = \frac{n!}{(1-\lambda)^{n+1}};$$

hence, $E(x^n) = M_x^{(n)}(0) = n!$. ∎

Drill Problem 3.1.1. *The RV x has PMF shown in Fig. 3.2. Find (a) $E(x)$, (b) $E(x^2)$, and (c) $E((x-2.125)^2)$.*

Answers: $\dfrac{199}{64}, \dfrac{61}{8}, \dfrac{17}{8}$.

Drill Problem 3.1.2. *We given $E(x) = 2.5$ and $E(y) = 10$. Determine: (a) $E(3x+4)$, (b) $E(x+y)$, and (c) $E(3x+8y+5)$.*

Answers: 12.5, 92.5, 11.5.

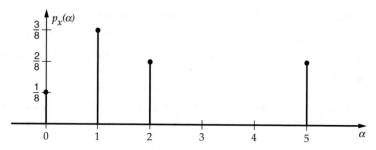

FIGURE 3.2: PMF for Drill Problem 3.1.1.

Drill Problem 3.1.3. *The PDF for the RV x is*

$$f_x(\alpha) = \begin{cases} \frac{3}{8}(\sqrt{\alpha} + \frac{1}{\sqrt{\alpha}}), & 0 < \alpha < 1 \\ 0, & \text{elsewhere.} \end{cases}$$

Find (a) $E(x)$, and (b) σ_x^2.

Answers: $\dfrac{17}{175}$, $\dfrac{2}{5}$.

Drill Problem 3.1.4. *The RV x has variance σ_x^2. Define the RVs y and z as $y = x + b$, and $z = ax$, where a and b are real constants. Find σ_y^2 and σ_z^2.*

Answers: σ_x^2, $a^2\sigma_x^2$.

Drill Problem 3.1.5. *The RV x has PDF $f_x(\alpha) = \frac{1}{2}e^{-|\alpha|}$. Find (a) $M_x(\lambda)$, (b) η_x, and (c) σ_x^2.*

Answers: 2; 0; $(1 - \lambda^2)^{-1}$, for $|\lambda| < 1$.

3.2   BOUNDS ON PROBABILITIES

In practice, one often has good estimates of some moments of a RV without having knowledge of the CDF. In this section, we investigate some important inequalities which enable one to establish bounds on probabilities which can be used when the CDF is not known. These bounds are also useful for gaining a "feel" for the information about the CDF contained in various moments.

Theorem 3.2.1.   (Generalized Chebyshev Inequality) *Let x be a RV on $(S, \Im, P)$, and let $\psi : \Re^* \mapsto \Re^*$ be strictly positive, even, nondecreasing on $(0, \infty]$, with $E(\psi(x)) < \infty$. Then for each $x_0 > 0$:*

$$P(|x(\zeta)| \geq x_0) \leq \frac{E(\psi(x))}{\psi(x_0)}. \tag{3.25}$$

Proof. Let $x_0 > 0$. Then

$$E(\psi(x)) = \int_{-\infty}^{\infty} \psi(\alpha) \, dF_x(\alpha)$$

$$= \int_{|\alpha|<x_0} \psi(\alpha) dF_x(\alpha) + \int_{|\alpha|\geq x_0} \psi(\alpha) dF_x(\alpha)$$

$$= \int_{|\alpha|\geq x_0} \psi(\alpha) dF_x(\alpha)$$

$$\geq \psi(x_0) \int_{|\alpha|\geq x_0} dF_x(\alpha)$$

$$= \psi(x_0) P(|x(\zeta)| \geq x_0).$$

∎

Corollary 3.2.1. (Markov Inequality) *Let x be a RV on (S, F, P), $x_0 > 0$, and $r > 0$. Then*

$$P(|x(\zeta)| \geq x_0) \leq \frac{E(|x(\zeta)|^r)}{x_0^r}. \tag{3.26}$$

Proof. The result follows from Theorem 1 with $\psi(x) = |x|^r$. ∎

Corollary 3.2.2. (Chebyshev Inequality) *Let x be a RV on $(S, \Im, P)$ with standard deviation σ_x, and let $\alpha > 0$. Then*

$$P(|x(\zeta) - \eta_x| \geq \alpha \sigma_x) \leq \frac{1}{\alpha^2}. \tag{3.27}$$

Proof. The desired result follows by applying the Markov Inequality to the RV $x - \eta_x$ with $r = 2$ and $x_0 = \alpha \sigma_x$. ∎

Example 3.2.1. *Random variable x has a mean and a variance of four, but an otherwise unknown CDF. Determine a lower bound on $P(|x - 4| < 8)$ using the Chebyshev Inequality.*

Solution. We have

$$P(|x - 4| \geq 8) = P(|x - \eta_x| \geq 4\sigma_x) \leq \frac{1}{16}.$$

Consequently,

$$P(|x - 4| < 8) = 1 - P(|x - 4| \geq 8) \geq 1 - \frac{1}{16} = \frac{15}{16}.$$

∎

Theorem 3.2.2. (Chernoff Bound) *Let x be a RV and assume both $M_x(\lambda)$ and $M_x(-\lambda)$ exist for some $\lambda > 0$, where M_x is the moment generating function for x. Then for any real x_0 we have*

$$P(x > x_0) \leq e^{-\lambda x_0} M_x(\lambda) \tag{3.28}$$

and

$$P(x \leq x_0) \leq e^{\lambda x_0} M_x(-\lambda). \tag{3.29}$$

The variable λ (which can depend on x_0) may be chosen to optimize the above bounds.

Proof. Noting that $e^{-\lambda(x_0 - \alpha)} \geq 1$ for $x_0 \leq \alpha$ we obtain

$$e^{-\lambda x_0} M_x(\lambda) = \int_{-\infty}^{\infty} e^{-\lambda(x_0 - \alpha)} \, dF_x(\alpha)$$

$$\geq \int_{x_0}^{\infty} dF_x(\alpha)$$

$$= P(x > x_0).$$

Similarly, since $e^{\lambda(x_0 - \alpha)} \geq 1$ for $x_0 \geq \alpha$ we obtain

$$e^{\lambda x_0} M_x(-\lambda) = \int_{-\infty}^{\infty} e^{\lambda(x_0 - \alpha)} \, dF_x(\alpha)$$

$$\geq \int_{-\infty}^{x_0} dF_x(\alpha)$$

$$= P(x \leq x_0).$$

■

Example 3.2.2. *The RV x has PDF $f_x(\alpha) = e^{-\alpha} u(\alpha)$. Compute bounds using the Markov Inequality, the Chebyshev Inequality, and the Chernoff Bound. Compare the bounds with corresponding quantities computed from the PDF.*

Solution. From Example 3.1.7 we have $E(x^n) = E(|x|^n) = n!$ and $M_x(\lambda) = (1 - \lambda)^{-1}$, for $\lambda < 1$. Consequently, $\sigma_x^2 = 2 - 1 = 1$.

Applying the Markov Inequality, we have

$$P(|x| \geq x_0) \leq \frac{n!}{x_0^n}, \qquad x_0 > 0.$$

For $x_0 = 10$, the upper bound is $0.1, 0.02, 3.63 \times 10^{-4}$ for $n = 1, 2$, and 10, respectively. Increasing n past x_0 results in a poorer upper bound for this example. Direct computation yields

$$P(|x| \geq x_0) = e^{-x_0}, \qquad x_0 > 0,$$

so that $P(|x| \geq 10) = e^{-10} = 4.54 \times 10^{-5}$.

Applying the Chebyshev Inequality,

$$P(|x - 1| \geq \alpha) \leq \frac{1}{\alpha^2}, \qquad \alpha > 0;$$

for $\alpha = 10$, the upper bound is 0.01. Direct computation yields (for $\alpha \geq 1$)

$$P(|x - 1| \geq \alpha) = \int_{1+\alpha}^{\infty} e^{-\alpha'} \, d\alpha = e^{-1-\alpha},$$

so that $P(|x - 1| \geq 10) = e^{-11} = 1.67 \times 10^{-5}$.

Applying the Chernoff Bound, we find (for any x_0)

$$P(x > x_0) \leq \frac{e^{-\lambda x_0}}{1 - \lambda}, \qquad 0 < \lambda < 1,$$

and

$$F_x(x_0) = P(x \leq x_0) \leq \frac{e^{\lambda x_0}}{1 + \lambda}, \qquad \lambda > 0.$$

The upper bound on $F_x(x_0)$ can be made arbitrarily small for $x_0 < 0$ by choosing a large enough λ. The Chernoff Bound thus allows us to conclude that $F_x(x_0) = 0$ for $x_0 < 0$. For $x_0 > 0$, let

$$g(\lambda) = \frac{e^{-\lambda x_0}}{1 - \lambda}.$$

Note that $g^{(1)}(\lambda) = 0$ for $\lambda = \lambda_0 = (x_0 - 1)/x_0$. Furthermore, $g^{(1)}(\lambda) > 0$ for $\lambda > \lambda_0$ and $g^{(1)}(\lambda) < 0$ for $\lambda < \lambda_0$. Hence, $\lambda = \lambda_0$ minimizes $g(\lambda)$, and we conclude that

$$P(x > x_0) \leq g(\lambda_0) = x_0 e^{1-x_0}, \qquad x_0 > 0.$$

For $x_0 = 10$, this upper bound yields 1.23×10^{-3}. Direct computation yields $P(x > x_0) = e^{-x_0} = 4.54 \times 10^{-5}$. ∎

Drill Problem 3.2.1. *Random variable x has $\eta_x = 7, \sigma_x = 4$, and otherwise unknown CDF. Using the Chebyshev inequality, determine a lower bound for (a) $P(-1 < x < 15)$, and (b) $P(-5 < x < 19)$.*

Answers: $\dfrac{3}{4}, \dfrac{8}{9}$.

Drill Problem 3.2.2. *Random variable x has an unknown PDF. How small should σ_x be to ensure that*

$$P(|x - \eta_x| < 1) \geq \frac{15}{16}?$$

Answer: $\sigma_x < 1/4$.

3.3   CHARACTERISTIC FUNCTION

Up to now, we have primarily described the uncertainty associated with a random variable using the PDF or CDF. In some applications, these functions may not be easy to work with. In this section, we introduce the use of transform methods in our study of random variables. Transforms provide another method of analysis that often yields more tractable solutions. Transforms also provide an alternate description of the probability distribution essential in our later study of linear systems.

Definition 3.3.1. *Let x be a RV on $(S, \Im, P)$. The **characteristic function for the RV** x is defined by*

$$\phi_x(t) = E(e^{jtx}) = \int_{-\infty}^{\infty} e^{jt\alpha} \, dF_x(\alpha), \qquad (3.30)$$

where $j^2 = -1$, and t is real.

Note the similarity of the characteristic function and the moment generating function. The characteristic function definition uses a complex exponential:

$$e^{jt\alpha} = \cos(t\alpha) + j \sin(t\alpha).$$

Note that since both t and α are real,

$$|e^{jt\alpha}|^2 = (e^{jt\alpha})(e^{jt\alpha})^* = e^{jt\alpha} e^{-jt\alpha} = 1.$$

If $z = x + jy$, where x and y are both real, then

$$e^z = e^x e^{jy} = e^x(\cos(y) + j \sin(y)),$$

so that $|e^z| = e^x$. Hence, $|e^z| \to +\infty$ as $x \to +\infty$ and $|e^z| \to 0$ as $x \to -\infty$.

Example 3.3.1. *(a) Find the characteristic function $\phi_x(t)$ for the RV x having CDF*

$$F_x(\alpha) = \sum_{i=1}^{n} a_i u(\alpha - \alpha_i),$$

where $a_i > 0, i = 1, 2, \ldots, n,$ and

$$\sum_{i=1}^{n} a_i = 1.$$

(b) Find $\phi_x(t)$ if the RV x has PDF $f_x(\alpha) = e^{-\alpha}u(\alpha)$.

(c) Find $\phi_x(t)$ if the RV x has PDF $f_x(\alpha) = e^{\alpha}u(-\alpha)$.

(d) Find $\phi_x(t)$ if the RV x has PDF $f_x(\alpha) = \frac{1}{2}e^{-|\alpha|}$.

(e) Find $\phi_x(t)$ if the RV x has PMF

$$p_x(\alpha) = \begin{cases} \dfrac{e^{-\eta}\eta^{\alpha}}{\alpha!}, & \alpha = 0, 1, \ldots \\ 0, & \text{otherwise.} \end{cases}$$

Solution. (a) We have

$$\phi_x(t) = \sum_{i=1}^{n} a_i \int_{-\infty}^{\infty} e^{j\alpha t} du(\alpha - \alpha_i) = \sum_{i=1}^{n} a_i e^{j\alpha_i t}.$$

Consequently, we know that any RV having a characteristic function of the form

$$\phi_x(t) = \sum_{i=1}^{n} a_i e^{j\alpha_i t}$$

is a discrete RV with CDF

$$F_x(\alpha) = \sum_{i=1}^{n} a_i u(\alpha - \alpha_i),$$

a PDF

$$f_x(\alpha) = \sum_{i=1}^{n} a_i \delta(\alpha - \alpha_i),$$

and a PMF

$$p_x(\alpha) = \begin{cases} a_i, & \alpha = \alpha_i, i = 1, 2, \ldots, n \\ 0, & \text{otherwise.} \end{cases}$$

(b) We have

$$\phi_x(t) = \int_{0}^{\infty} e^{\alpha(-1+jt)} d\alpha = \frac{1}{1 - jt}.$$

(c) We have

$$\phi_x(t) = \int_{-\infty}^{0} e^{\alpha(1+jt)} \, d\alpha = \frac{1}{1-jt}.$$

(d) The given PDF may be expressed as

$$f_x(\alpha) = \frac{1}{2}(e^{\alpha}u(-\alpha) + e^{-\alpha}u(\alpha))$$

so that we can use (b) and (c) to obtain

$$\phi_x(t) = \frac{1}{2}\left(\frac{1}{1+jt} + \frac{1}{1-jt}\right) = \frac{1}{1+t^2}.$$

(e) We have

$$\begin{aligned}
\phi_x(t) &= \sum_{k=0}^{\infty} \frac{e^{jkt}e^{-\eta}\eta^k}{k!} \\
&= e^{-\eta} \sum_{k=0}^{\infty} \frac{e^{(jt_\eta)k}}{k!} \\
&= e^{-\eta} \exp(e^{jt}\eta) \\
&= \exp(\eta(e^{jt} - 1)).
\end{aligned}$$ ■

The characteristic function is an integral transform in which there is a unique one-to-one relationship between the probability density function and the characteristic function. For each PDF f_x there is only one corresponding ϕ_x. We often find one from the other from memory or from transform tables—the preceding example provides the results for several important cases.

Unlike the moment generating function, the characteristic function always exists. Like the moment generating function, the characteristic function is often used to compute moments for a random variable.

Theorem 3.3.1. *The characteristic function $\phi_x(t)$ always exists and satisfies*

$$|\phi_x(t)| \leq 1. \tag{3.31}$$

Proof. Since $|e^{jt\alpha}| = 1$ for all real t and all real α we have

$$|\phi_x(t)| \leq \int_{-\infty}^{\infty} |e^{jt\alpha}| \, dF_x(\alpha) = 1.$$ ■

Theorem 3.3.2. (Moment Generating Property) *Let*

$$\phi_x^{(n)}(t) = \frac{d^n \phi_x(t)}{dt^n} \qquad (3.32)$$

and assume that $\phi_x^{(n)}(0)$ *exists. Then*

$$E(x^n) = (-j)^n \phi_x^{(n)}(0). \qquad (3.33)$$

Proof. We have

$$\phi_x^{(n)}(t) = E\left(\frac{d^n e^{jtx}}{dt^n}\right) = E((jx)^n e^{jtx})$$

from which the desired result follows by letting $t = 0$. ∎

Example 3.3.2. *The RV x has the Bernoulli PMF*

$$p_x(k) = \begin{cases} \binom{n}{k} p^k q^{n-k}, & k = 0, 1, \ldots, n \\ 0, & \text{otherwise,} \end{cases}$$

where $0 \le q = 1 - p \le 1$. *Find the characteristic function* $\phi_x(t)$ *and use it to find* $E(x)$ *and* σ_x^2.

Solution. Applying the Binomial Theorem, we have

$$\phi_x(t) = \sum_{k=0}^{n} \binom{n}{k} (e^{jt} p)^k q^{n-k} = (pe^{jt} + q)^n.$$

Then

$$\phi_x^{(1)}(t) = n(pe^{jt} + q)^{n-1} jpe^{jt},$$

and

$$\phi_x^{(2)}(t) = n(n-1)(pe^{jt} + q)^{n-2}(jpe^{jt})^2 + n(pe^{jt} + q)^{n-1} j^2 pe^{jt},$$

so that $\phi_x^{(1)}(0) = jnp$ and $\phi_x^{(2)}(0) = -n^2 p^2 + np^2 - np = -n^2 p^2 - npq$. Hence, $E(x) = np$ and $E(x^2) = n^2 p^2 + npq$. Finally, $\sigma_x^2 = E(x^2) - E^2(x) = npq$. ∎

Lemma 3.3.1. *Let the RV* $y = ax + b$, *where a and b are constants and the RV x has characteristic function* $\phi_x(t)$. *Then the characteristic function for y is*

$$\phi_y(t) = e^{jbt} \phi_x(at). \qquad (3.34)$$

Proof. By definition

$$\phi_y(t) = E(e^{jyt}) = E(e^{j(ax+b)t}) = e^{jbt} E(e^{jx(at)}).$$

∎

Lemma 3.3.2. *Let the RV $y = ax + b$. Then if $a > 0$*

$$F_y(\alpha) = F_x((\alpha - b)/a). \qquad (3.35)$$

If $a < 0$ then

$$F_y(\alpha) = 1 - F_x(((\alpha - b)/a)^-). \qquad (3.36)$$

Proof. With $a > 0$,

$$F_y(\alpha) = P(ax + b \le \alpha) = P(x \le (\alpha - b)/a).$$

With $a < 0$,

$$F_y(\alpha) = P(x \ge (\alpha - b)/a).$$

∎

Let the discrete RV x be a lattice RV with $p_k = P(x(\zeta) = a + kh)$ and

$$\sum_{k=-\infty}^{\infty} p_k = 1. \qquad (3.37)$$

Then

$$\phi_x(t) = e^{jat} \sum_{k=-\infty}^{\infty} p_k e^{jkht}. \qquad (3.38)$$

Note that

$$|\phi_x(t)| = \left| \sum_{k=-\infty}^{\infty} p_k e^{jkht} \right|. \qquad (3.39)$$

Since

$$e^{jkh(t+\tau)} = e^{jkht}(e^{jh\tau})^k = e^{jkht} \qquad (3.40)$$

for $\tau = 2\pi/h$, we find that $|\phi_x(t + \tau)| = |\phi_x(t)|$; i.e., $|\phi_x(t)|$ is periodic in t with period $\tau = 2\pi/h$. We may interpret p_k as the kth complex Fourier series coefficient for $e^{-jat}\phi_x(t)$. Hence, p_k can be determined from ϕ_x using

$$p_k = \frac{h}{2\pi} \int_{-\pi/h}^{\pi/h} \phi_x(t) e^{-jat} e^{-jkht} \, dt. \qquad (3.41)$$

An expansion of the form (3.38) is unique: If ϕ_x can be expressed as in (3.38) then the parameters a and h as well as the coefficients $\{p_k\}$ can be found by inspection, and the RV x is known to be a discrete lattice RV.

Example 3.3.3. *Let the RV x have characteristic function*

$$\phi_x(t) = e^{j4t} \cos(5t).$$

Find the PMF $p_x(\alpha)$.

Solution. Using Euler's identity

$$\phi_x(t) = e^{j4t} \left(\frac{1}{2} e^{-j5t} + \frac{1}{2} e^{j5t} \right) = e^{jat} \sum_{k=-\infty}^{\infty} p_k e^{jkht}.$$

We conclude that $a = 4$, $h = 5$, and $p_{-1} = p_1 = 0.5$, so that

$$p_x(\alpha) = \begin{cases} 0.5, & \alpha = 4 - 5 = -1, \quad \alpha = 4 + 5 = 9 \\ 0, & \text{otherwise.} \end{cases}$$

∎

Example 3.3.4. *The RV x has characteristic function*

$$\phi_x(t) = \frac{0.1 e^{j0.5t}}{1 - 0.9 e^{j3t}}.$$

Show that x is a discrete lattice RV and find the PMF p_x.

Solution. Using the sum of a geometric series, we find

$$\phi_x(t) = 0.1 e^{j0.5t} \sum_{k=0}^{\infty} (0.9 e^{j3t})^k.$$

Comparing this with (3.38) we find $a = 0.5$, $h = 3$, and

$$p_x(0.5 + 3k) = p_k = \begin{cases} 0.1(0.9)^k, & k = 0, 1, \dots \\ 0, & \text{otherwise.} \end{cases}$$

∎

The characteristic function $\phi_x(t)$ is (within a factor of 2π) the inverse Fourier transform of the PDF $f_x(\alpha)$. Consequently, the PDF can be obtained from the characteristic function via a Fourier transform operation. In many applications, the CDF is the required function. With the aid of the following lemma, we establish below that the CDF may be obtained "directly" from the characteristic function.

Lemma 3.3.3. *Define*

$$S(\beta, T) = \frac{1}{\pi} \int_{-T}^{T} \frac{e^{j\beta t}}{jt} dt. \tag{3.42}$$

Then

$$S(\beta, T) = \frac{2}{\pi} \int_0^T \frac{sin(\beta t)}{t} dt. \tag{3.43}$$

and

$$\lim_{T \to \infty} S(\beta, T) = \begin{cases} -1 & \text{if } \beta < 0 \\ 0 & \text{if } \beta = 0 \\ 1 & \text{if } \beta > 0. \end{cases} \tag{3.44}$$

Proof.   We have

$$S(\beta, T) = \frac{1}{\pi} \int_{-T}^0 \frac{e^{j\beta t}}{jt} dt + \frac{1}{\pi} \int_0^T \frac{e^{j\beta t}}{jt} dt$$

$$= \frac{1}{\pi} \int_0^T \frac{e^{-j\beta \tau}}{-j\tau} d\tau + \frac{1}{\pi} \int_0^T \frac{e^{j\beta t}}{jt} dt$$

$$= \frac{2}{\pi} \int_0^T \frac{sin(\beta t)}{t} dt$$

$$= \frac{2}{\pi} \int_0^{\beta T} \frac{sin(\tau)}{\tau} d\tau .$$

The desired result follows by using the fact that

$$\int_0^\infty \frac{sin\, t}{t} dt = \frac{\pi}{2},$$

and noting that $S(-\beta, T) = -S(\beta, T)$. ∎

Theorem 3.3.3. *Let ϕ_x be the characteristic function for the RV x with CDF F_x, and assume $F_x(\alpha)$ is continuous at $\alpha = a$ and $\alpha = b$. Then if $b > a$ we have*

$$F_x(b) - F_x(a) = \lim_{T \to \infty} \frac{1}{2\pi} \int_{-T}^T \frac{e^{-jat} - e^{-jbt}}{jt} \phi_x(t) \, dt . \tag{3.45}$$

Proof. Let

$$I(T) = \frac{1}{2\pi} \int\limits_{-T}^{T} \frac{e^{-jat} - e^{-jbt}}{jt} \phi_x(t) \, dt.$$

From the definition of a characteristic function

$$I(T) = \frac{1}{2\pi} \int\limits_{-T}^{T} \left(\int\limits_{-\infty}^{\infty} \frac{e^{-jat} - e^{-jbt}}{jt} e^{jt\alpha} \, dF_x(\alpha) \right) dt.$$

Interchanging the order of integration we have

$$I(T) = \frac{1}{2} \int\limits_{-\infty}^{\infty} (S(\alpha - a, T) - S(\alpha - b, T)) \, dF_x(\alpha).$$

Interchanging the order of the limit and integration we have

$$\lim_{T \to \infty} I(T) = \int\limits_{a}^{b} dF_x(\alpha) = F_x(b) - F_x(a).$$

∎

Corollary 3.3.1. *Assume the RV x has PDF f_x. Then*

$$f_x(\alpha) = \lim_{T \to \infty} \frac{1}{2\pi} \int\limits_{-T}^{T} \phi_x(t) e^{-j\alpha t} \, dt. \qquad (3.46)$$

Proof. The desired result follows from the above theorem by letting $b = \alpha$, $a = \alpha - h$, and $h > 0$. Then

$$f_x(\alpha) = \lim_{h \to 0} \frac{F_x(\alpha) - F_x(\alpha - h)}{h}$$

$$= \lim_{T \to \infty} \frac{1}{2\pi} \int\limits_{-T}^{T} \lim_{h \to 0} \frac{e^{jht} - 1}{jth} e^{-j\alpha t} \phi_x(t) \, dt.$$

∎

In some applications, a closed form for the characteristic function is available but the inversion integrals for obtaining either the CDF or the PDF cannot be obtained analytically. In these cases, a numerical integration may be performed efficiently by making use of the FFT (fast Fourier transform) algorithm.

The relationship between the PDF $f_x(\alpha)$ and the characteristic function $\phi_x(t)$ is that of a Fourier transform pair. Although several definitions of a Fourier transform exist, we present below the commonly used definition within the field of Electrical Engineering.

Definition 3.3.2. *We define the **Fourier transform** of a function $g(t)$ by*

$$G(\omega) = \mathcal{F}\{g(t)\} = \int_{-\infty}^{\infty} g(t)e^{-j\omega t}\,dt. \tag{3.47}$$

*The corresponding **inverse Fourier transform** of $G(\omega)$ is defined by*

$$g(t) = \mathcal{F}^{-1}\{G(\omega)\} = \frac{1}{2\pi}\int_{-\infty}^{\infty} G(\omega)e^{j\omega t}\,d\omega. \tag{3.48}$$

If $g(t)$ is absolutely integrable; i.e., if

$$\int_{-\infty}^{\infty} |g(t)|\,dt < \infty,$$

then $G(\omega)$ exists and the inverse Fourier transform integral converges to $g(t)$ for all t where $g(t)$ is continuous. The preceding development for characteristic functions can be used to justify this Fourier transform result. In particular, we note that

$$\int_{-\infty}^{\infty} g(t)e^{-j\omega t}\,dt$$

should be interpreted as

$$\lim_{T\to\infty} \int_{-T}^{T} g(t)e^{-j\omega t}\,dt.$$

Using these definitions, we find that

$$\phi_x(t) = 2\pi\mathcal{F}^{-1}\{f_x(\alpha)\} = \int_{-\infty}^{\infty} f_x(\alpha)e^{j\alpha t}\,d\alpha, \tag{3.49}$$

and

$$f_x(\alpha) = \frac{1}{2\pi}\mathcal{F}\{\phi_x(t)\} = \frac{1}{2\pi}\int_{-\infty}^{\infty} \phi_x(t)e^{-j\alpha t}\,dt. \tag{3.50}$$

The Fourier transform $G(\omega) = \mathcal{F}\{g(t)\}$ is unique; i.e., if $G(\omega) = \mathcal{F}\{g(t)\}$, then we know that $g(t) = \mathcal{F}^{-1}\{G(\omega)\}$ for almost all values of t. The same is true for characteristic functions.

Drill Problem 3.3.1. *Random variable x has PDF*

$$f_x(\alpha) = 0.5(u(\alpha + 1) - u(\alpha - 1)).$$

Find: (a) $\phi_x(0)$, (b) $\phi_x(\pi/4)$, (c) $\phi_x(\pi/2)$, and (d) $\phi_x(\pi)$.

Answers: $1, \dfrac{2}{\pi}, \dfrac{\sin(\pi/4)}{\pi/4}, \quad 0.$

Drill Problem 3.3.2. *The PDF for RV x is $f_x(\alpha) = e^{-\alpha}u(\alpha)$. Use the characteristic function to obtain: (a) $E(x)$, (b) $E(x^2)$, (c) σ_x, and (d) $E(x^3)$.*

Answers: 2, 1, 6, 1.

3.4 CONDITIONAL EXPECTATION

Definition 3.4.1. *The **conditional expectation** $g(x)$, given event A, is defined by*

$$E(g(x)|A) = \int_{-\infty}^{\infty} g(\alpha)\, dF_{x|A}(\alpha \mid A). \tag{3.51}$$

*The **conditional mean and conditional variance of the RV** x, given event A, are similarly defined as*

$$\eta_{x|A} = E(x \mid A) \tag{3.52}$$

and

$$\sigma_{x|A}^2 = E((x - \eta_{x|A})^2 \mid A) = E(x^2 \mid A) - \eta_{x|A}^2. \tag{3.53}$$

*Similarly, the **conditional characteristic function of the RV** x, given event A, is defined as*

$$\phi_{x|A}(t|A) = E(e^{jxt} \mid A) = \int_{-\infty}^{\infty} e^{j\alpha t}\, dF_{x|A}(\alpha \mid A). \tag{3.54}$$

Example 3.4.1. *An urn contains four red balls and three blue balls. Three balls are drawn without replacement from the urn. Let A denote the event that at least two red balls are selected, and let RV x denote the number of red balls selected. Find $E(x)$ and $E(x \mid A)$.*

Solution. Let R_i denote a red ball drawn on the ith draw, and B_i denote a blue ball. Since x is the number of red balls, x can only take on the values 0,1,2,3. The sequence event $B_1 B_2 B_3$ occurs with probability 1/35; hence $P(x = 0) = 1/35$. Next, consider the sequence event $R_1 B_2 B_3$ which occurs with probability 4/35. Since there are three sequence events which contain one red ball, we have $P(x = 1) = 12/35$. Similarly, $P(x = 2) = 18/35$ and $P(x = 3) = 4/35$. We thus find that

$$E(x) = 0 \cdot \frac{1}{35} + 1 \cdot \frac{12}{35} + 2 \cdot \frac{18}{35} + 3 \cdot \frac{4}{35} = \frac{12}{7}.$$

Now, $P(A) = P(x \geq 2) = 22/35$ so that

$$p_{x|A}(\alpha \mid A) = \begin{cases} \dfrac{18/35}{22/35} = \dfrac{9}{11}, & \alpha = 2 \\ \dfrac{4/35}{22/35} = \dfrac{2}{11}, & \alpha = 3 \\ 0, & \text{otherwise.} \end{cases}$$

Consequently,

$$E(x \mid A) = 2 \cdot \frac{9}{11} + 3 \cdot \frac{2}{11} = \frac{24}{11}. \qquad \blacksquare$$

Example 3.4.2. *Find the conditional mean and conditional variance for the RV x, given event* $A = \{x > 1\}$, *where* $f_x(\alpha) = e^{-\alpha} u(\alpha)$.

Solution. First, we find

$$P(A) = \int_1^\infty f_x(\alpha) \, d\alpha = \int_1^\infty e^{-\alpha} \, d\alpha = e^{-1}.$$

Then $f_{x|A}(\alpha \mid A) = e^{1-\alpha} u(\alpha - 1)$. The conditional mean and conditional variance, given A, can be found using $f_{x|A}$ using integration by parts. Here, we use the characteristic function method. The conditional characteristic function is

$$\phi_{x|A}(t \mid A) = \frac{1}{P(A)} \int_1^\infty e^{\alpha(-1+jt)} \, d\alpha = \frac{e^{jt}}{1 - jt}.$$

Differentiating, we find

$$\phi_{x|A}^{(1)}(t \mid A) = je^{jt} \left(\frac{1}{1 - jt} + \frac{1}{(1 - jt)^2} \right),$$

so that $\phi_{x|A}^{(1)}(0 \mid A) = j2$ and

$$\phi_{x|A}^{(2)}(t \mid A) = -e^{jt} \left(\frac{1}{1 - jt} + \frac{1}{(1 - jt)^2} \right) + je^{jt} \left(\frac{j}{(1 - jt)^2} + \frac{2j}{(1 - jt)^3} \right)$$

so that $\phi_{x|A}^{(2)}(0 \mid A) = -5$. Thus $\eta_{x|A} = -j(j2) = 2$ and $\sigma_{x|A}^2 = (-j)^2(-5) - 2^2 = 1$. $\blacksquare$

Drill Problem 3.4.1. *The RV x has PMF shown in Fig. 3.2 . Event $A = \{x \leq 3\}$. Find (a) $\eta_{x|A}$ and (b) $\sigma^2_{x|A}$.*

Answers: 17/36, 7/6.

Drill Problem 3.4.2. *Random variable x has PDF*

$$f_x(\alpha) = \frac{3}{8}(\sqrt{\alpha} + \frac{1}{\sqrt{\alpha}})(u(\alpha) - u(\alpha - 1)).$$

Event $A = \{x < 0.25\}$. Find (a) $E(3x + 2 \mid A)$ and (b) $\sigma^2_{x|A}$.

Answers: 8879/1537900, 589/260.

3.5 SUMMARY

In this chapter, the statistical expectation operation is defined and used to determine bounds on probabilities.

The mean (or expected value) of the RV x is defined as

$$\eta_x = E(x) = \int_{-\infty}^{\infty} \alpha \, dF_x(\alpha) \tag{3.55}$$

and the variance of x as $\sigma^2_x = E((x - \eta_x)^2)$.

Expectation is a linear operation, the expected value of a constant is the constant.

The moment generating function (when it exists) is defined as $M_x(\lambda) = E(e^{\lambda x})$, from which moments can be computed as $E(x^n) = M_x^{(n)}(0)$.

Partial knowledge about a CDF for a RV x is contained in the moments for x. In general, knowledge of all moments for x is not sufficient to determine the CDF F_x. However, available moments can be used to compute bounds on probabilities. In particular, the probability that a RV x deviates from its mean by at least $\alpha \times \sigma$ is upper bounded by $1/\alpha^2$. Tighter bounds generally require more information about the CDF—higher order moments, for example.

The characteristic function $\phi_x(t) = E(e^{jtx})$ is related to the inverse Fourier transform of the PDF f_x. All information concerning a CDF F_x is contained in the characteristic function ϕ_x. In particular, the CDF itself can be obtained from the characteristic function.

Conditional expectation, given an event, is a linear operation defined in terms of the conditional CDF:

$$E(g(x)|A) = \int_{-\infty}^{\infty} g(\alpha) \, dF_{x|A}(\alpha \mid A). \tag{3.56}$$

Conditional moments and the conditional characteristic function are similarly defined.

3.6   PROBLEMS

1. The sample space is $S = \{a_1, a_2, a_3, a_4, a_5\}$ with probabilities $P(a_1) = 0.15$, $P(a_2) = 0.2$, $P(a_3) = 0.1$, $P(a_4) = 0.25$, and $P(a_5) = 0.3$. Random variable x is defined as $x(a_i) = 2i - 1$. Find: (a) η_x, (b) $E(x^2)$.

2. Consider a department in which all of its graduate students range in age from 22 to 28. Additionally, it is three times as likely a student's age is from 22 to 24 as from 25 to 28. Assume equal probabilities within each age group. Let random variable x equal the age of a graduate student in this department. Determine: (a) $E(x)$, (b) $E(x^2)$, (c) σ_x.

3. A class contains five students of about equal ability. The probability a student obtains an A is 1/5, a B is 2/5, and a C is 2/5. Let random variable x equal the number of students who earn an A in the class. Determine: (a) $p_x(\alpha)$, (b) $E(x)$, (c) σ_x.

4. Random variable x has the following PDF

$$f_x(\alpha) = \begin{cases} 0.5(\alpha + 1), & -1 < \alpha < 1 \\ 0, & \text{otherwise.} \end{cases}$$

Determine: (a) $E(x)$, (b) σ_x^2, (c) $E(1/(x + 1))$, (d) $\sigma_{1/(x+1)}^2$.

5. The PDF for random variable y is

$$f_y(y_o) = \begin{cases} \sin(y_o), & 0 < y_o < \pi/2 \\ 0, & \text{otherwise,} \end{cases}$$

and $g(y) = \sin(y)$. Determine $E(g(y))$.

6. Sketch these PDF's, and, for each, find the variance of x: (a) $f_x(\alpha) = 0.5e^{-|\alpha|}$, (b) $f_x(\alpha) = 5e^{-10|\alpha|}$.

7. The grade distribution for Professor S. Rensselaer's class in probability theory is shown in Fig. 3.3. (a) Write a mathematical expression for $f_x(\alpha)$. (b) Determine $E(x)$. (c) Suppose grades are assigned on the basis of: 90–100 = A = 4 honor points, 75–90 = B = 3 honor points, 60–75 = C = 2 honor points, 55–60 = D = 1 honor point, and 0–55 = F = 0 honor points. Find the honor points PDF. (d) Find the honor points average.

8. A PDF is given by

$$f_x(\alpha) = \frac{1}{2}\delta(\alpha + 1.5) + \frac{1}{8}\delta(\alpha) + \frac{3}{8}\delta(\alpha - 2).$$

Determine: (a) $E(x)$, (b) σ_x^2.

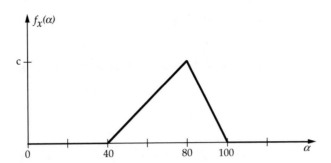

FIGURE 3.3: Probability density function for Problem 7.

9. A PDF is given by

$$f_x(\alpha) = \frac{1}{5}\delta(\alpha + 1) + \frac{2}{5}\delta(\alpha) + \frac{3}{10}\delta(\alpha - 1) + \frac{1}{10}\delta(\alpha - 2).$$

Determine: (a) $E(x)$, (b) $E(x^2)$.

10. A mixed random variable has a CDF given by

$$F_x(\alpha) = \begin{cases} 0, & \alpha < 0 \\ \alpha/4, & 0 \le \alpha < 1 \\ 1 - e^{-0.6931\alpha}, & 1 \le \alpha. \end{cases}$$

Determine: (a) $E(x)$, (b) σ_x^2.

11. A mixed random variable has a PDF given by

$$f_x(\alpha) = \frac{1}{4}\delta(\alpha + 1) + \frac{3}{8}\delta(\alpha - 1) + \frac{1}{4}(u(\alpha + 1) - u(\alpha - 0.5)).$$

Determine: (a) $E(x)$, (b) σ_x^2.

12. Let RV x have mean η_x and variance σ_x^2. (a) Show that

$$E(|x - a|^2) = \sigma_x^2 + (\eta_x - a)^2$$

for any real constant a. (b) Find a so that $E(|x - a|^2)$ is minimized.

13. The random variable y has $\eta_y = 10$ and $\sigma_y^2 = 2$. Find (a) $E(y^2)$ and (b) $E((y - 3)^2)$.

14. The **median** for a RV x is the value of α for which $F_x(\alpha) = 0.5$. Let x be a RV with median m. (a) Show that for any real constant a:

$$E(|x - a|) = E(|x - m|) + 2 \int_a^m (\alpha - a) \, dF_x(\alpha).$$

(b) Find the constant a for which $E(|x - a|)$ is minimized.

15. Use integration by parts to show that

$$E(x) = \int_0^\infty (1 - F_x(\alpha))\, d\alpha - \int_{-\infty}^0 F_x(\alpha)\, d\alpha.$$

16. Show that

$$E(|x|) = \int_{-\infty}^0 F_x(\alpha)\, d\alpha + \int_0^\infty (1 - F_x(\alpha))\, d\alpha.$$

17. Random variable x has $\eta_x = 50$, $\sigma_x = 5$, and an otherwise unknown CDF. Using the Chebyshev Inequality, find a lower bound on $P(30 < x < 70)$.

18. Suppose random variable x has a mean of 6 and a variance of 25. Using the Chebyshev Inequality, find a lower bound on $P(|x - 6| < 50)$.

19. RV x has a mean of 20 and a variance of 4. Find an upper bound on $P(|x - 20| \geq 8)$.

20. Random variable x has an unknown PDF. How small should σ_x be so that $P(|x - \eta_x| \geq 2) \leq 1/9$?

21. RVs x and y have PDFs f_x and f_y, respectively. Show that

$$E(\ln f_x(x)) \geq E(\ln f_y(x)).$$

22. Find the characteristic function for random variable x if

$$p_x(\alpha) = \begin{cases} p, & \alpha = 1 \\ q, & \alpha = 0 \\ 0, & \text{otherwise.} \end{cases}$$

23. RV x has PDF $f_x(\alpha) = u(\alpha) - u(\alpha - 1)$. Determine: (a) ϕ_x. Use the characteristic function to find: (b) $E(x)$, (c) $E(x^2)$, (d) σ_x.

24. Random variable x has PDF $f_x(\alpha) = 3e^{3\alpha} u(-\alpha)$. Find ϕ_x.

25. Show that the characteristic function for a Cauchy random variable with PDF

$$f_x(\alpha) = \frac{1}{\pi(1 + \alpha^2)}$$

is $\phi_x(t) = e^{-|t|}$.

26. Given $f_x(\alpha) = 0.5\beta \exp(-\beta|\alpha|)$. Find (a) ϕ_x. Use ϕ_x to determine: (b) $E(x)$, (c) $E(x^2)$, and (d) σ_x.

27. Random variable x has the PDF $f_x(\alpha) = 2\alpha(u(\alpha) - u(\alpha - 1))$. (a) Find ϕ_x. (b) Show that $\phi_x(0) = 1$. (c) Find $E(x)$ using the characteristic function.

28. Suppose

$$F_x(\alpha) = \begin{cases} 1, & 0 \le \alpha \\ \exp(3\alpha), & \alpha < 0. \end{cases}$$

Use the characteristic function to determine: (a) $E(x)$, (b) $E(x^2)$, (c) $E(x^3)$, and (d) σ_x^2.

29. Suppose x is a random variable with

$$p_x(\alpha) = \begin{cases} \beta\gamma^\alpha, & \alpha = 0, 1, 2, \ldots \\ 0, & \text{otherwise.} \end{cases}$$

where β and γ are constants, and $0 < \gamma < 1$. As a function of γ, determine: (a) β, (b) $M_x(\lambda)$, (c) $\phi_x(t)$, (d) $E(x)$, (e) σ_x^2.

30. RV x has characteristic function

$$\phi_x(t) = (pe^{jt} + (1 - p))^n,$$

where $0 < p < 1$. Find the PMF $p_x(\alpha)$.

31. The PDF for RV x is $f_x(\alpha) = \alpha e^{-\alpha} u(\alpha)$. Find (a) ϕ_x, (b) η_x, and (c) σ_x^2.

32. RV x has characteristic function

$$\phi_x(t) = \begin{cases} 1 - \frac{|t|}{a}, & |t| < a \\ 0, & \text{otherwise.} \end{cases}$$

Find the PDF f_x.

33. RV x has PDF

$$f_x(\alpha) = \begin{cases} c\left(1 - \frac{|\alpha|}{a}\right), & |\alpha| < a \\ 0, & \text{otherwise.} \end{cases}$$

Find the constant c and find the characteristic function ϕ_x.

34. The random variable x has PMF

$$p_x(\alpha) = \begin{cases} 2/13, & \alpha = -1 \\ 3/13, & \alpha = 1 \\ 4/13, & \alpha = 2 \\ 3/13, & \alpha = 3 \\ 1/13, & \alpha = 4 \\ 0, & \text{otherwise.} \end{cases}$$

Random variable $z = 3x + 2$ and event $A = \{x > 2\}$. Find (a) $E(x)$, (b) $E(x|A)$, (c) $E(z)$, (d) σ_z^2.

35. The head football coach at the renowned Fargo Polytechnic Institute is in serious trouble. His job security is directly related to the number of football games the team wins each year. The team has lost its first three games in the eight game schedule. The coach knows that if the team loses five games, he will be fired immediately. The alumni hate losing and consider a tie as bad as a loss. Let x be a random variable whose value equals the number of games the present head coach wins. Assume the probability of winning any game is 0.6 and independent of the results of other games. Determine: (a) $E(x)$, (b) σ_x, (c) $E(x|x > 3)$, (d) $\sigma_{x|x>3}^2$.

36. Consider Problem 35. The team loves the head coach and does not want to lose him. The more desperate the situation becomes for the coach, the better the team plays. Assume the probability the team wins a game is dependent on the total number of losses as $P(W|L) = 0.2L$, where W is the event the team wins a game and L is the total number of losses for the team. Let A be the event the present head coach is fired before the last game of the season. Determine: (a) $E(x)$, (b) σ_x, (c) $E(x|A)$.

37. Random variable y has the PMF

$$
p_y(\alpha) = \begin{cases}
1/8, & \alpha = 0 \\
3/16, & \alpha = 1 \\
1/4, & \alpha = 2 \\
5/16, & \alpha = 3 \\
1/8, & \alpha = 4 \\
0, & \text{otherwise.}
\end{cases}
$$

Random variable $w = (y - 2)^2$ and event $A = \{y \geq 2\}$. Determine: (a) $E(y)$, (b) $E(y \mid A)$, (c) $E(w)$.

38. In BME Bioinstrumentation lab, each student is given one transistor to use during one experiment. The probability a student destroys a transistor during this experiment is 0.7. Let random variable x equal the number of destroyed transistors. In a class of five students, determine: (a) $E(x)$, (b) σ_x, (c) $E(x \mid x < 4)$, (d) $\sigma_{x|x<4}$.

39. Consider Problem 38. Transistors cost 20 cents each plus one dollar for mailing (all transistors). Let random variable z equal the amount of money in dollars that is spent on new transistors for the class of five students. Determine: (a) $p_z(\alpha)$, (b) $F_z(\alpha)$, (c) $E(z)$, (d) σ_z.

40. An urn contains ten balls with labels 1, 2, 2, 3, 3, 3, 5, 5, 7, and 8. A ball is drawn at random. Let random variable x be the number printed on the ball and event $A = \{x \text{ is odd}\}$. Determine: (a) $E(x)$, (b) $E(x^2)$, (c) σ_x, (d) $E(5x - 2)$, (e) σ_{3x}, (f) $E(5x - 3x^2)$, (g) $E(x \mid A)$, (h) $E(x^2 \mid A)$, (i) $E(3x^2 - 2x \mid A)$.

41. A biased four-sided die, with faces labeled 1, 2, 3 and 4, is tossed once. If the number which appears is odd, the die is tossed again. Let random variable x equal the sum of numbers which appear if the die is tossed twice or the number which appears on the first toss if it is only thrown once. The die is biased so that the probability of a particular face is proportional to the number on that face. Event $A = \{\text{first die toss number is odd}\}$ and $B = \{\text{second die toss number is odd}\}$. Determine: (a) $p_x(\alpha)$, (b) $E(x)$, (c) $E(x \mid B)$, (d) σ_x^2, (e) $\sigma_{x \mid B}^2$, (f) whether events A and B are independent.

42. Suppose the following information is known about random variable x. First, the values x takes on are a subset of integers. Additionally, $F_x(-1) = 0$, $F_x(3) = 5/8$, $F_x(6) = 1$, $p_x(0) = 1/8$, $p_x(1) = 1/4$, $p_x(6) = 1/8$, $E(x) = 47/16$, and $E(x \mid x > 4) = 16/3$. Determine (a) $p_x(\alpha)$, (b) $F_x(\alpha)$, (c) σ_x^2, (d) $\sigma_{x \mid x > 4}^2$.

43. A biased pentahedral die, with faces labeled 1, 2, 3, 4, and 5, is tossed once. The die is biased so that the probability of a particular face is proportional to the number on that face. Let x be a random variable whose values equal the number which appears on the tossed die. The outcome of the die toss determines which of five biased coins is flipped. The probability a head appears for the ith coin is $1/(6 - i)$, $i = 1, 2, 3, 4, 5$. Define event $A = \{x \text{ is even}\}$ and event $B = \{\text{tail appears}\}$. Determine: (a) $E(x)$, (b) σ_x, (c) $E(x \mid B)$, (d) $\sigma_{x \mid B}^2$, (e) whether events A and B are independent.

44. Given

$$F_x(\alpha) = \begin{cases} 0, & \alpha < 0 \\ 3(\alpha - \alpha^2 + \alpha^3/3), & 0 \le \alpha < 1 \\ 1, & 1 \le \alpha, \end{cases}$$

and event $A = \{1/4 < x\}$. Determine: (a) $E(x)$, (b) $E(x^2)$, (c) $E(5x^2 - 3x + 2)$, (d) $E(4x^2 - 4)$, (e) $E(3x + 2 \mid A)$, (f) $E(x^2 \mid A)$, (g) $E(3x^2 - 2x + 2 \mid A)$.

45. The PDF for random variable x is

$$f_x(\alpha) = \begin{cases} 1/\alpha, & 1 < \alpha < 2.7183 \\ 0, & \text{otherwise,} \end{cases}$$

and event $A = \{x < 1.6487\}$. Determine: (a) $E(x)$, (b) σ_x^2, (c) $E(x \mid A)$, (d) $\sigma_{x \mid A}^2$.

46. With the PDF for random variable x given by

$$f_x(\alpha) = \begin{cases} \dfrac{4}{\pi(1+\alpha^2)}, & 0 < \alpha < 1 \\ 0, & \text{otherwise,} \end{cases}$$

determine: (a) $E(x)$; (b) $E(x|x > 1/8)$; (c) $E(2x - 1)$; (d) $E(2x - 1 | x > 1/8)$; (e) the variance of x; (f) the variance of x, given $x > 1/8$.

47. A random variable x has CDF

$$F_x(\alpha) = \left(\alpha + \frac{1}{2}\right) u\left(\alpha + \frac{1}{2}\right) - \alpha u(\alpha) + \frac{1}{4}\alpha u(\alpha - 1) + \left(\frac{1}{2} - \frac{\alpha}{4}\right) u(\alpha - 2),$$

and event $A = \{x \geq 1\}$. Find: (a) $E(x)$, (b) σ_x^2, (c) $E(x|A)$, and (d) $\sigma_{x|A}^2$.

CHAPTER 4

Bivariate Random Variables

In many situations, we must consider models of probabilistic phenomena which involve more than one random variable. These models enable us to examine the interaction among variables associated with the underlying experiment. For example, in studying the performance of a telemedicine system, variables such as cosmic radiation, sun spot activity, solar wind, and receiver thermal noise might be important noise level attributes of the received signal. The experiment is modeled with n random variables. Each outcome in the sample space is mapped by the n RVs to a point in real n-dimensional Euclidean space.

In this chapter, the joint probability distribution for two random variables is considered. The joint CDF, joint PMF, and joint PDF are first considered, followed by a discussion of two–dimensional Riemann-Stieltjes integration. The previous chapter demonstrated that statistical expectation can be used to bound event probabilities; this concept is extended to the two-dimensional case in this chapter. The more general case of n-dimensional random variables is treated in a later chapter.

4.1   BIVARIATE CDF

Definition 4.1.1. *A **two-dimensional (or bivariate) random variable** $\mathbf{z} = (x, y)$ defined on a probability space $(S, \Im, P)$ is a mapping from the outcome space S to $\Re^* \times \Re^*$; i.e., to each outcome $\zeta \in S$ corresponds a pair of real numbers, $\mathbf{z}(\zeta) = (x(\zeta), y(\zeta))$. The functions x and y are required to be random variables. Note that $\mathbf{z} : S \mapsto \Re^* \times \Re^*$, and that we need $z^{-1}([-\infty, \alpha] \times [-\infty, \beta]) \in \Im$ for all real α and β.*

The two-dimensional mapping performed by the bivariate RV z is illustrated in Fig. 4.1.

Definition 4.1.2. *The **joint CDF (or bivariate cumulative distribution function)** for the RVs x and y (both of which are defined on the same probability space$(S, \Im, P)$) is defined by*

$$F_{x,y}(\alpha, \beta) = P(\{\zeta \in S : x(\zeta) \le \alpha, y(\zeta) \le \beta\}). \qquad (4.1)$$

Note that $F_{x,y} : \Re^* \times \Re^* \mapsto [0, 1]$. With $A = \{\zeta \in S : x(\zeta) \le \alpha\}$ and $B = \{\zeta \in S : y(\zeta) \le \beta\}$, the joint CDF is given by $F_{x,y}(\alpha, \beta) = P(A \cap B)$.

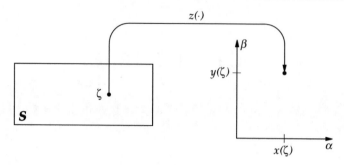

FIGURE 4.1: A bivariate random variable $z(\cdot)$ maps each outcome in S to a pair of extended real numbers.

Using the relative frequency approach to probability assignment, a bivariate CDF can be estimated as follows. Suppose that the RVs x and y take on the values x_i and y_i on the ith trial of an experiment, with $i = 1, 2, \ldots, n$. The **empirical distribution function**

$$\hat{F}_{x,y}(\alpha, \beta) = \frac{1}{n} \sum_{i=1}^{n} u(\alpha - x_i)u(\beta - y_i) \qquad (4.2)$$

is an estimate of the CDF $F_{x,y}(\alpha, \beta)$, where $u(\cdot)$ is the unit step function. Note that $\hat{F}_{x,y}(\alpha, \beta) = n(\alpha, \beta)/n$, where $n(\alpha, \beta)$ is the number of observed pairs (x_i, y_i) satisfying $x_i \le \alpha$, $y_i \le \beta$.

Example 4.1.1. *The bivariate RV $\mathbf{z} = (x, y)$ is equally likely to take on the values $(1, 2)$, $(1, 3)$, and $(2, 1)$. Find the joint CDF $F_{x,y}$.*

Solution. Define the region of $\Re^* \times \Re^*$:

$$A(\alpha, \beta) = \{(\alpha', \beta') : \alpha' \le \alpha, \beta' \le \beta\},$$

and note that

$$F_{x,y}(\alpha, \beta) = P((x, y) \in A(\alpha, \beta)).$$

We begin by placing a dot in the $\alpha' - \beta'$ plane for each possible value of (x, y), as shown in Fig. 4.2(a). For $\alpha < 1$ or $\beta < 1$ there are no dots inside $A(\alpha, \beta)$ so that $F_{x,y}(\alpha, \beta) = 0$ in this region. For $1 \le \alpha < 2$ and $2 \le \beta < 3$, only the dot at $(1, 2)$ is inside $A(\alpha, \beta)$ so that $F_{x,y}(\alpha, \beta) = 1/3$ in this region. Continuing in this manner, the values of $F_{x,y}$ shown in Fig. 4.2(b) are easily obtained. Note that $F_{x,y}(\alpha, \beta)$ can only increase or remain constant as either α or β is increased. ∎

Theorem 4.1.1. (Properties of Joint CDF) *The joint CDF $F_{x,y}$ satisfies:*

(i) $F_{x,y}(\alpha, \beta)$ is monotone nondecreasing in each of the variables α and β,

(ii) $F_{x,y}(\alpha, \beta)$ is right-continuous in each of the variables α and β,

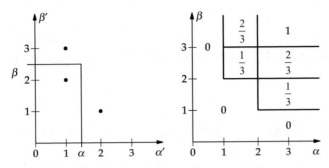

FIGURE 4.2: Possible values and CDF representation for Example 4.1.1.

(iii) $F_{x,y}(-\infty, \beta) = F_{x,y}(\alpha, -\infty) = F_{x,y}(-\infty, -\infty) = 0$,

(iv) $F_{x,y}(\alpha, \infty) = F_x(\alpha)$, $F_{x,y}(\infty, \beta) = F_y(\beta)$, $F_{x,y}(\infty, \infty) = 1$. *The CDFs F_x and F_y are called the **marginal CDFs** for x and y, respectively.*

Proof. (i) With $\alpha_2 > \alpha_1$ we have

$$\{x \leq \alpha_2, y \leq \beta_1\} = \{x \leq \alpha_1, y \leq \beta_1\} \cup \{\alpha_1 < x \leq \alpha_2, y \leq \beta_1\}.$$

Since

$$\{x \leq \alpha_1, y \leq \beta_1\} \cap \{\alpha_1 < x \leq \alpha_2, y \leq \beta_1\} = \emptyset,$$

we have

$$F_{x,y}(\alpha_2, \beta_1) = F_{x,y}(\alpha_1, \beta_1) + P(\zeta \in \{\alpha_1 < x \leq \alpha_2, y \leq \beta_1\})$$
$$\geq F_{x,y}(\alpha_1, \beta_1).$$

Similarly, with $\beta_2 > \beta_1$ we have

$$\{x \leq \alpha_1, y \leq \beta_2\} = \{x \leq \alpha_1, y \leq \beta_1\} \cup \{x \leq \alpha_1, \beta_1 < y \leq \beta_2\}.$$

Since

$$\{x \leq \alpha_1, y \leq \beta_1\} \cap \{x \leq \alpha_1, \beta_1 < y \leq \beta_2\} = \emptyset,$$

we have

$$F_{x,y}(\alpha_1, \beta_2) = F_{x,y}(\alpha_1, \beta_1) + P(\zeta \in \{x \leq \alpha_1, \beta_1 < y \leq \beta_2\})$$
$$\geq F_{x,y}(\alpha_1, \beta_1).$$

(ii) follows from the above proof of (i) by taking the limit (from the right) as $\alpha_2 \to \alpha_1$ and $\beta_2 \to \beta_1$.

(iii) We have

$$\{\zeta \in S : x(\zeta) = -\infty, y(\zeta) \le \beta\} \subset \{\zeta \in S : x(\zeta) = -\infty\}$$

and

$$\{\zeta \in S : x(\zeta) \le \alpha, y(\zeta) = -\infty\} \subset \{\zeta \in S : y(\zeta) = -\infty\};$$

result (iii) follows by noting that from the definition of a RV, $P(x(\zeta) = -\infty) = P(y(\zeta) = -\infty) = 0$.

(iv) We have

$$F_{x,y}(\alpha, \infty) = P(\{\zeta : x(\zeta) \le \alpha\} \cap S) = P(x(\zeta) \le \alpha) = F_x(\alpha).$$

Similarly, $F_{x,y}(\infty, \beta) = F_y(\beta)$, and $F_{x,y}(\infty, \infty) = 1$. ∎

Probabilities for rectangular-shaped events in the x, y plane can be obtained from the bivariate CDF in a straightforward manner. Define the left-sided **difference operators** Δ_1 and Δ_2 by

$$\Delta_1(h) F_{x,y}(\alpha, \beta) = F_{x,y}(\alpha, \beta) - F_{x,y}(\alpha - h, \beta), \tag{4.3}$$

and

$$\Delta_2(h) F_{x,y}(\alpha, \beta) = F_{x,y}(\alpha, \beta) - F_{x,y}(\alpha, \beta - h), \tag{4.4}$$

with $h > 0$. Then, with $h_1 > 0$ and $h_2 > 0$ we have

$$\begin{aligned}
\Delta_2(h_2)\Delta_1(h_1) F_{x,y}(\alpha, \beta) &= F_{x,y}(\alpha, \beta) - (F_{x,y}(\alpha - h_1, \beta) - (F_{x,y}(\alpha, \beta - h_2) \\
&\quad - F_{x,y}(\alpha - h_1, \beta - h_2)) \\
&= P(\alpha - h_1 < x \le \alpha, y \le \beta) - P(\alpha - h_1 < x \le \alpha, y \le \beta - h_2) \\
&= P(\alpha - h_1 < x(\zeta) \le \alpha, \beta - h_2 < y(\zeta) \le \beta).
\end{aligned} \tag{4.5}$$

With $a_1 < b_1$ and $a_2 < b_2$ we thus have

$$\begin{aligned}
P(a_1 < x \le b_1, a_2 < y \le b_2) &= \Delta_2(b_2 - a_2)\Delta_1(b_1 - a_1) F_{x,y}(b_1, b_2) \\
&= F_{x,y}(b_1, b_2) - F_{x,y}(a_1, b_2) \\
&\quad - (F_{x,y}(b_1, a_2) - F_{x,y}(a_1, a_2)).
\end{aligned} \tag{4.6}$$

Example 4.1.2. *The RVs x and y have joint CDF*

$$F_{x,y}(\alpha, \beta) = \begin{cases} 0, & \alpha < 0 \\ 0, & \beta < 0 \\ 0.5\alpha\beta, & 0 \le \alpha < 1, \quad 0 \le \beta < 1 \\ 0.5\beta, & 1 \le \alpha < 2, \quad 0 \le \beta < 1 \\ 0.25 + 0.5\beta, & 2 \le \alpha, \quad 0 \le \beta < 1 \\ 0.5\alpha, & 0 \le \alpha < 1, \quad 1 \le \beta \\ 0.5, & 1 \le \alpha < 2, \quad 1 \le \beta \\ 0.75, & 2 \le \alpha < 3, \quad 1 \le \beta \\ 1, & 3 \le \alpha, \quad 1 \le \beta. \end{cases}$$

Find: (a) $P(x = 2, y = 0)$, (b) $P(x = 3, y = 1)$, (c) $P(0.5 < x < 2, 0.25 < y \le 3)$, (d) $P(0.5 < x \le 1, 0.25 < y \le 1)$.

Solution. We begin by using two convenient methods for representing the bivariate CDF graphically. The first method simply divides the $\alpha - \beta$ plane into regions with the functional relationship (or value) for the CDF written in the appropriate region to represent the height of the CDF above the region. The results are shown in Fig. 4.3. The second technique is to plot a family of curves for $F_{x,y}(\alpha, \beta)$ vs. α for various ranges of β. Such a family of curves for this example is shown in Fig. 4.4.

(a) We have

$$\begin{aligned} P(x = 2, y = 0) &= P(2^- < x \le 2, 0^- < y \le 0) \\ &= \Delta_2(0^+)\Delta_1(0^+)F_{x,y}(2, 0) \\ &= F_{x,y}(2, 0) - F_{x,y}(2^-, 0) - (F_{x,y}(2, 0^-) - F_{x,y}(2^-, 0^-)) \\ &= 0.25. \end{aligned}$$

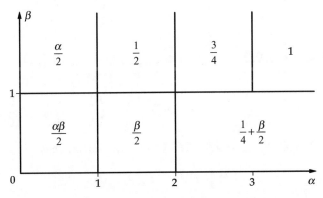

FIGURE 4.3: Two-dimensional representation of bivariate CDF for Example 4.1.2.

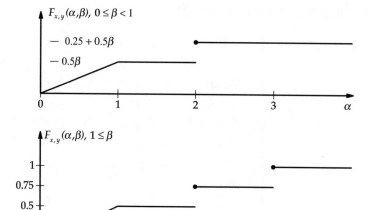

FIGURE 4.4: Bivariate CDF for Example 4.1.2.

(b) Proceeding as above

$$
\begin{aligned}
P(x = 3, y = 1) &= \Delta_2(0^+)\Delta_2(0^+)F_{x,y}(3, 1) \\
&= F_{x,y}(3, 1) - F_{x,y}(3^-, 1) - (F_{x,y}(3, 1^-) - F_{x,y}(3^-, 1^-)) \\
&= 1 - 0.75 - (0.75 - 0.75) = 0.25.
\end{aligned}
$$

(c) We have

$$
\begin{aligned}
P(0.5 < x < 2, 0.25 < y \le 3) &= F_{x,y}(2^-, 3) - F_{x,y}(0.5, 3) \\
&\quad - (F_{x,y}(2^-, 0.25) - F_{x,y}(0.5, 0.25)) \\
&= \frac{1}{2} - \frac{1}{4} - \left(\frac{1}{8} - \frac{1}{2}\frac{1}{2}\frac{1}{4}\right) = \frac{3}{16}.
\end{aligned}
$$

(d) As above, we have

$$
\begin{aligned}
P(0.5 < x \le 1, 0.25 < y \le 1) &= F_{x,y}(1, 1) - F_{x,y}(0.5, 1) \\
&\quad - (F_{x,y}(1, 0.25) - F_{x,y}(0.5, 0.25)) \\
&= \frac{1}{2} - \frac{1}{4} - \left(\frac{1}{8} - \frac{1}{16}\right) = \frac{3}{16}.
\end{aligned}
$$

∎

Definition 4.1.3. *The jointly distributed RVs x and y are* **independent**

$$
F_{x,y}(\alpha, \beta) = F_x(\alpha)F_y(\beta) \tag{4.7}
$$

for all real values of α and β.

In Chapter 1, we defined the two events A and B to be independent iff $P(A \cap B) = P(A)P(B)$. With $A = \{\zeta \in S : x(\zeta) \le \alpha\}$ and $B = \{\zeta \in S : y(\zeta) \le \beta\}$, the RVs x and y are independent iff A and B are independent for all real values of α and β. In many applications, physical arguments justify an assumption of independence. When used, an independence assumption greatly simplifies the analysis. When not fully justified, however, the resulting analysis is highly suspect—extensive testing is then needed to establish confidence in the simplified model.

Note that **if x and y are independent** then for $a_1 < b_1$ and $a_2 < b_2$ we have

$$P(a_1 < x \le b_1, a_2 < y \le b_2) = F_{x,y}(b_1, b_2) - F_{x,y}(a_1, b_2) - (F_{x,y}(b_1, a_2) - F_{x,y}(a_1, a_2))$$
$$= (F_x(b_1) - F_x(a_1))(F_y(b_2) - F_y(a_2)). \tag{4.8}$$

4.1.1 Discrete Bivariate Random Variables

Definition 4.1.4. *The bivariate RV (x, y) defined on the probability space $(S, \Im, P)$ is **bivariate discrete** if the joint CDF $F_{x,y}$ is a jump function; i.e., iff there exists a countable set $D_{x,y} \subset \Re \times \Re$ such that*

$$P(\{\zeta \in S : (x(\zeta), y(\zeta)) \in D_{x,y}\}) = 1. \tag{4.9}$$

*In this case, we also say that the RVs x and y are **jointly discrete**. The function*

$$p_{x,y}(\alpha, \beta) = P(x = \alpha, y = \beta) \tag{4.10}$$

*is called the **bivariate probability mass function or simply the joint PMF for the jointly distributed discrete RVs** x and y. We will on occasion refer to the set $D_{x,y}$ as the support set for the PMF $p_{x,y}$. The support set for the PMF $p_{x,y}$ is the set of points for which $p_{x,y}(\alpha, \beta) \ne 0$.*

Theorem 4.1.2. *The bivariate PMF $p_{x,y}$ can be found from the joint CDF as*

$$p_{x,y}(\alpha, \beta) = \lim_{h_2 \to 0} \lim_{h_1 \to 0} \Delta_2(h_2)\Delta_1(h_1)F_{x,y}(\alpha, \beta) \tag{4.11}$$
$$= F_{x,y}(\alpha, \beta) - F_{x,y}(\alpha^-, \beta) - (F_{x,y}(\alpha, \beta^-) - F_{x,y}(\alpha^-, \beta^-)),$$

where the limits are through positive values of h_1 and h_2. Conversely, the joint CDF $F_{x,y}$ can be found from the PMF $p_{x,y}$ as

$$F_{x,y}(\alpha, \beta) = \sum_{\beta' \le \beta} \sum_{\alpha' \le \alpha} p_{x,y}(\alpha', \beta'). \tag{4.12}$$

The probability that the bivariate discrete RV $(x, y) \in A$ can be computed using

$$P((x, y) \in A) = \sum_{(\alpha, \beta) \in A} p_{x,y}(\alpha, \beta). \tag{4.13}$$

All summation indices are assumed to be in the support set for $p_{x,y}$.

Proof. The theorem is a direct application of the bivariate CDF and the definition of a PMF.   ∎

Any function $p_{x,y}$ mapping $\Re^* \times \Re^*$ to $\Re \times \Re$ with a discrete support set $D_{x,y} = D_x \times D_y$ and satisfying

$$p_{x,y}(\alpha, \beta) \geq 0 \quad \text{for all real } \alpha \text{ and } \beta, \tag{4.14}$$

$$p_x(\alpha) = \sum_{\beta \in D_y} p_{x,y}(\alpha, \beta), \tag{4.15}$$

and

$$p_y(\beta) = \sum_{\alpha \in D_x} p_{x,y}(\alpha, \beta), \tag{4.16}$$

where p_x and p_y are valid one-dimensional PMFs, is a legitimate bivariate PMF.

Corollary 4.1.1. *The **marginal PMFs** p_x and p_y may be obtained from the bivariate PMF as*

$$p_x(\alpha) = \sum_{\beta} p_{x,y}(\alpha, \beta) \tag{4.17}$$

and

$$p_y(\beta) = \sum_{\alpha} p_{x,y}(\alpha, \beta). \tag{4.18}$$

Theorem 4.1.3. *The jointly discrete RVs x and y are independent iff*

$$p_{x,y}(\alpha, \beta) = p_x(\alpha) p_y(\beta) \tag{4.19}$$

for all real α and β.

Proof. The theorem follows from the definition of PMF and independence.   ∎

Example 4.1.3. *The RVs x and y have joint PMF specified in the table below.*

α	β	$p_{x,y}(\alpha, \beta)$
−1	0	1/8
−1	1	1/8
0	3	1/8
1	−1	2/8
1	1	1/8
2	1	1/8
3	3	1/8

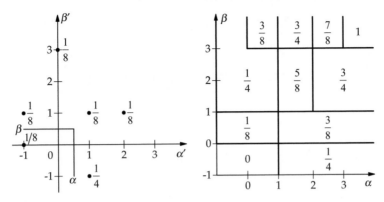

FIGURE 4.5: PMF and CDF representations for Example 4.1.3.

(a) Sketch the two-dimensional representations for the PMF and the CDF. (b) Find p_x. (c) Find p_y. (d) Find $P(x < y)$. (e) Are x and y independent?

Solution. (a) From the previous table, the two–dimensional representation for the PMF shown in Fig. 4.5(a) is easily obtained. Using the sketch for the PMF, visualizing the movement of the (α, β) values and summing all PMF weights below and to the left of (α, β), the two-dimensional representation of the CDF shown in Fig. 4.5(b) is obtained.
(b) We have

$$p_x(\alpha) = \sum_\beta p_{x,y}(\alpha, \beta),$$

so that

$$p_x(-1) = p_{x,y}(-1, 0) + p_{x,y}(-1, 1) = 2/8,$$

$$p_x(0) = p_{x,y}(0, 3) = 1/8,$$

$$p_x(1) = p_{x,y}(1, -1) + p_{x,y}(1, 1) = 3/8,$$

$$p_x(2) = p_{x,y}(2, 1) = 1/8,$$

$$p_x(3) = p_{x,y}(3, 3) = 1/8.$$

(c) Proceeding as in part (b),

$$p_y(-1) = p_{x,y}(1, -1) = 2/8,$$

$$p_y(0) = p_{x,y}(-1, 0) = 1/8,$$

$$p_y(1) = p_{x,y}(-1, 1) + p_{x,y}(1, 1) + p_{x,y}(2, 1) = 3/8,$$

$$p_y(3) = p_{x,y}(0, 3) + p_{x,y}(3, 3) = 2/8.$$

(d) We have

$$P(x < y) = p_{x,y}(-1, 0) + p_{x,y}(-1, 1) + p_{x,y}(0, 3) = 3/8.$$

(e) Since $p_{x,y}(1, 1) = 1/8 \neq p_x(1)p_y(1) = 9/64$, we find that x and y are not independent. ∎

Example 4.1.4. *The jointly discrete RVs x and y have joint PMF*

$$p_{x,y}(k, \ell) = \begin{cases} c\gamma^k\lambda^{|k-\ell|}, & k, \ell \text{ nonnegative integers} \\ 0, & \text{otherwise}, \end{cases}$$

where $0 < \gamma < 1$, and $0 < \lambda < 1$. Find: (a) the marginal PMF p_x, (b) the constant c, (c) $P(x < y)$.

Solution. (a) For $k = 0, 1, \ldots,$

$$p_x(k) = \sum_{\ell=-\infty}^{\infty} p_{x,y}(k, \ell)$$

$$= c\gamma^k \left(\lambda^k \sum_{\ell=0}^{k} \lambda^{-\ell} + \lambda^{-k} \sum_{\ell=k+1}^{\infty} \lambda^{\ell} \right)$$

$$= c\gamma^k \left(\frac{1 - \lambda^{k+1}}{1 - \lambda} + \frac{\lambda}{1 - \lambda} \right)$$

$$= \frac{c\gamma^k(1 + \lambda - \lambda^{k+1})}{1 - \lambda}.$$

(b) We have

$$1 = \sum_{k=0}^{\infty} p_x(k) = \frac{c}{1 - \lambda} \left(\frac{1 + \lambda}{1 - \lambda} - \frac{\lambda}{1 - \lambda\gamma} \right).$$

so that

$$c = \frac{(1 - \lambda)(1 - \gamma)(1 - \lambda\gamma)}{1 - \lambda^2\gamma}.$$

(c) We find

$$P(x < y) = \sum_{k=0}^{\infty} \sum_{\ell=k+1}^{\infty} c \left(\frac{\gamma}{\lambda}\right)^k \lambda^\ell$$

$$= c \sum_{k=0}^{\infty} \left(\frac{\gamma}{\lambda}\right)^k \frac{\lambda^{k+1}}{1-\lambda}$$

$$= \frac{c\lambda}{1-\lambda} \frac{1}{1-\lambda}$$

∎

4.1.2 Bivariate Continuous Random Variables

Definition 4.1.5. *A bivariate RV (x, y) defined on the probability space $(S, \Im, P)$ is **bivariate continuous** if the joint CDF $F_{x,y}$ is absolutely continuous. To avoid technicalities, we simply note that if $F_{x,y}$ is absolutely continuous then $F_{x,y}$ is continuous everywhere and $F_{x,y}$ is differentiable except perhaps at isolated points. Consequently, there exists a function $f_{x,y}$ satisfying*

$$F_{x,y}(\alpha, \beta) = \int_{-\infty}^{\beta} \int_{-\infty}^{\alpha} f_{x,y}(\alpha', \beta') d\alpha' d\beta' \qquad (4.20)$$

*The function $f_{x,y}$ is called the **bivariate probability density function** for the continuous RV (x, y), or simply the **joint PDF** for the RVs x and y.*

Theorem 4.1.4. *The joint PDF for the jointly distributed RVs x and y can be determined from the joint CDF as*

$$f_{x,y}(\alpha, \beta) = \frac{\partial^2 F_{x,y}(\alpha, \beta)}{\partial \beta \partial \alpha}$$

$$= \lim_{h_2 \to 0} \lim_{h_1 \to 0} \frac{\Delta_2(h_2)\Delta_1(h_1) F_{x,y}(\alpha, \beta)}{h_2 h_1}, \qquad (4.21)$$

where the limits are taken over positive values of h_1 and h_2, corresponding to a left-sided derivative in each coordinate.

The univariate, or marginal, PDFs f_x and f_y may be determined from the joint PDF $f_{x,y}$ as

$$f_x(\alpha) = \int_{-\infty}^{\infty} f_{x,y}(\alpha, \beta) d\beta, \qquad (4.22)$$

and

$$f_y(\beta) = \int_{-\infty}^{\infty} f_{x,y}(\alpha, \beta) d\alpha. \qquad (4.23)$$

Furthermore, we have

$$f_{x,y}(\alpha, \beta) \geq 0 \qquad (4.24)$$

and

$$\int\limits_{-\infty}^{\infty} \int\limits_{-\infty}^{\infty} f_{x,y}(\alpha, \beta)\, d\alpha d\beta = 1. \qquad (4.25)$$

The probability that $(x, y) \in A$ may be computed from

$$P((x, y) \in A) = \int_{(\alpha,\beta)\in A} f_{x,y}(\alpha, \beta) d\alpha d\beta. \qquad (4.26)$$

This integral represents the volume under the joint PDF surface above the region A.

Proof.   By definition,

$$f_x(\alpha) = \lim_{h \to 0} \frac{\Delta_1(h) F_{x,y}(\alpha, \infty)}{h}$$

$$= \lim_{h \to 0} \frac{1}{h} \int\limits_{-\infty}^{\infty} \int\limits_{\alpha-h}^{\alpha} f_{x,y}(\alpha', \beta)\, d\alpha'\, d\beta$$

$$= \int\limits_{-\infty}^{\infty} f_{x,y}(\alpha, \beta)\, d\beta.$$

The remaining conclusions of the theorem are straightforward consequences of the properties of a joint CDF and the definition of a joint PDF.   ■

We will often refer to the set of points where the joint PDF $f_{x,y}$ is nonzero as the **support set** for $f_{x,y}$. For jointly continuous RVs x and y, this support set is often called the **support region**. Letting $R_{x,y}$ denote the support region, for any event A we have

$$P(A) = P(A \cap R_{x,y}). \qquad (4.27)$$

Any function $f_{x,y}$ mapping $\Re^* \times \Re^*$ to $\Re \times \Re$ with a support set $R_{x,y} = R_x \times R_y$ and satisfying

$$f_{x,y}(\alpha, \beta) \geq 0 \quad \text{for (almost) all real } \alpha \text{ and } \beta, \qquad (4.28)$$

$$f_x(\alpha) = \int_{\beta \in R_y} f_{x,y}(\alpha, \beta)\, d\beta, \qquad (4.29)$$

and

$$f_y(\beta) = \int_{\alpha \in R_x} f_{x,y}(\alpha, \beta) \, d\beta, \qquad (4.30)$$

where f_x and f_y are valid one-dimensional PDFs, is a legitimate bivariate PDF.

Theorem 4.1.5. *The jointly continuous RVs x and y are independent iff*

$$f_{x,y}(\alpha, \beta) = f_x(\alpha) f_y(\beta) \qquad (4.31)$$

for all real α and β except perhaps at isolated points.

Proof. The theorem follows directly from the definition of joint PDF and independence. ∎

Example 4.1.5. *Let $A = \{(x, y) : -1 < x < 0.5, 0.25 < y < 0.5\}$, and*

$$f_{x,y}(\alpha, \beta) = \begin{cases} 4\alpha\beta, & 0 \le \alpha \le 1, 0 \le \beta \le 1 \\ 0, & \text{otherwise.} \end{cases}$$

Find: (a) $P(A)$, (b) f_x, (c) f_y. (d) Are x and y independent?

Solution. Note that the support region for $f_{x,y}$ is the unit square $R = \{(\alpha, \beta) : 0 < \alpha < 1, 0 < \beta < 1\}$. A three-dimensional plot of the PDF is shown in Fig. 4.6.

(a) Since A represents a rectangular region, we can find $P(A)$ from the joint CDF and (4.27) as

$$P(A) = P(A \cap R) = \Delta_2(0.5 - 0.25)\Delta_1(0.5 - 0)F_{x,y}(0.5^-, 0.5^-).$$

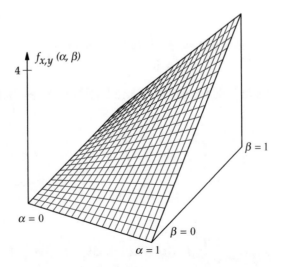

FIGURE 4.6: Three-dimensional plot of PDF for Example 4.1.5.

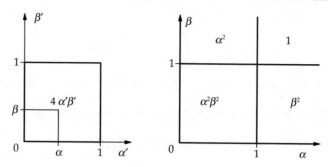

FIGURE 4.7: PDF and CDF representations for Example 4.1.5.

For $0 \le \alpha \le 1$ and $0 \le \beta \le 1$ we have

$$F_{x,y}(\alpha, \beta) = \int_0^\beta \int_0^\alpha 4\alpha' \beta' \, d\alpha' d\beta' = \alpha^2 \beta^2.$$

Substituting, we find

$$
\begin{aligned}
P(A) &= \Delta_2(0.25)(F_{x,y}(0.5, 0.5) - F_{x,y}(0, 0.5)) \\
&= F_{x,y}(0.5, 0.5) - F_{x,y}(0.5, 0.25) \\
&= \frac{3}{64}.
\end{aligned}
$$

Alternately, using the PDF directly, $P(A)$ is the volume under the PDF curve and above A:

$$P(A) = \int_{0.25}^{0.5} \int_0^{0.5} 4\alpha\beta \, d\alpha d\beta = \frac{3}{64}.$$

Two-dimensional representations for the PDF and CDF are shown in Fig. 4.7.
(b) We have

$$
f_x(\alpha) = \begin{cases} \int_0^1 f_{x,y}(\alpha, \beta) d\beta = 2\alpha, & 0 \le \alpha \le 1 \\ 0, & \text{otherwise.} \end{cases}
$$

(c) We have

$$
f_y(\beta) = \begin{cases} \int_0^1 f_{x,y}(\alpha, \beta) d\beta = 2\alpha, & 0 \le \alpha \le 1 \\ 0, & \text{otherwise.} \end{cases}
$$

(d) Since $f_{x,y}(\alpha, \beta) = f_x(\alpha)f_y(\beta)$ for all real α and β we find that the RVs x and y are independent. ∎

Example 4.1.6.   *The jointly distributed RVs x and y have joint PDF*

$$f_{x,y}(\alpha, \beta) = \begin{cases} 6(1 - \sqrt{\alpha/\beta}), & 0 \le \alpha \le \beta \le 1 \\ 0, & \text{otherwise,} \end{cases}$$

Find (a) $P(A)$, where $A = \{(x, y) : 0 < x < 0.5, 0 < y < 0.5\}$; (b) f_x; (c) f_y, and (d) $F_{x,y}$.

Solution.   (a) The support region R for the given PDF is

$$R = \{(\alpha, \beta) : 0 < \alpha < \beta < 1\}.$$

A two-dimensional representation for $f_{x,y}$ is shown in Fig. 4.8. Integrating with respect to α first,

$$P(A) = P(A \cap R) = \int_0^{0.5} \int_0^{\beta} 6(1 - \sqrt{\alpha/\beta})\, d\alpha d\beta = 2\int_0^{0.5} \beta\, d\beta = \frac{1}{4}.$$

One could integrate with respect to β first:

$$P(A) = P(A \cap R) = \int_0^{0.5} \int_{\alpha}^{0.5} 6(1 - \sqrt{\alpha/\beta})\, d\beta\, d\alpha.$$

This also provides the result—at the expense of a more difficult integration.

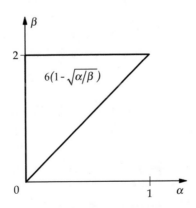

FIGURE 4.8: PDF representation for Example 4.1.6.

(b) For $0 < \alpha < 1$,

$$f_x(\alpha) = \int_\alpha^1 6(1 - \sqrt{\alpha/\beta}) \, d\beta = 6(1 - 2\sqrt{\alpha} + \alpha).$$

(c) For $0 < \beta < 1$,

$$f_y(\beta) = \int_0^\beta 6(1 - \sqrt{\alpha/\beta}) \, d\alpha = 2\beta.$$

(d) For $(\alpha, \beta) \in R_{x,y}$ (i.e., $0 \le \alpha \le \beta \le 1$),

$$F_{x,y}(\alpha, \beta) = 6 \int_0^\alpha \int_{\alpha'}^\beta (1 - \sqrt{\alpha'/\beta'}) \, d\beta' d\alpha'$$

$$= 6 \int_0^\alpha (\beta - 2\sqrt{\alpha'\beta} + \alpha') \, d\alpha'$$

$$= 6\alpha\beta - 8\alpha\sqrt{\alpha\beta} + 3\alpha^2.$$

For $0 \le \beta \le 1$ and $\beta \le \alpha$,

$$F_{x,y}(\alpha, \beta) = 6 \int_0^\beta \int_{\alpha'}^\beta (1 - \sqrt{\alpha'/\beta'}) \, d\beta' d\alpha'$$

$$= 6 \int_0^\beta (\beta - 2\sqrt{\alpha'\beta} + \alpha') \, d\alpha'$$

$$= \beta^2.$$

For $0 \le \alpha \le 1$ and $\beta \ge 1$),

$$F_{x,y}(\alpha, \beta) = 6 \int_0^\alpha \int_{\alpha'}^1 (1 - \sqrt{\alpha'/\beta'}) \, d\beta' d\alpha'$$

$$= 6 \int_0^\alpha (1 - 2\sqrt{\alpha'} + \alpha') \, d\alpha'$$

$$= 6\alpha - 8\alpha^{3/2} + 3\alpha^2.$$

■

4.1.3 Bivariate Mixed Random Variables

Definition 4.1.6. *The bivariate RV (x, y) defined on the probability space $(S, \Im, P)$ is a **mixed** RV if it is neither discrete nor continuous.*

Unlike the one-dimensional case, where the Lebesgue Decomposition Theorem enables us to separate a univariate CDF into discrete and continuous parts, the bivariate case requires either the two-dimensional Riemann-Stieltjes integral or the use of Dirac delta functions along with the two-dimensional Riemann integral. We illustrate the use of Dirac delta functions below. The two-dimensional Riemann-Stieltjes integral is treated in the following section. The probability that $(x, y) \in A$ can be expressed as

$$P((x, y) \in A) = \int_{(\alpha, \beta) \in A} dF_{x,y}(\alpha, \beta) = \int_A dF_{x,y}(\alpha, \beta). \qquad (4.32)$$

Example 4.1.7. *The RVs x and y have joint CDF*

$$F_{x,y}(\alpha, \beta) = \begin{cases} 0, & \alpha < 0 \\ 0, & \beta < 0 \\ \alpha\beta/4, & 0 \le \alpha < 1, \quad 0 \le \beta < 2 \\ \beta/4, & 1 \le \alpha, \quad 0 \le \beta < 2 \\ \alpha/2, & 0 \le \alpha < 1, \quad 2 \le \beta \\ 1, & 1 \le \alpha, \quad 2 \le \beta. \end{cases}$$

(a) Find an expression for $F_{x,y}$ using unit-step functions. (b) Find F_x and F_y. Are x and y independent? (c) Find f_x, f_y, and $f_{x,y}$ (using Dirac delta functions). (d) Evaluate

$$I = \int\limits_{-\infty}^{\infty} \int\limits_{-\infty}^{\beta} f_x(\alpha) f_y(\beta) \, d\alpha \, d\beta.$$

(e) Find $P(x \le y)$.

Solution. (a) A two-dimensional representation for the given CDF is illustrated in Fig. 4.9. This figure is useful for obtaining the CDF representation in terms of unit step functions. Using the figure, the given CDF can be expressed as

$$F_{x,y}(\alpha, \beta) = \frac{1}{4}(u(\alpha) - u(\alpha - 1))(\alpha\beta u(\beta) + (2\alpha - \alpha\beta)u(\beta - 2))$$
$$+ \frac{1}{4}u(\alpha - 1)(\beta u(\beta) + (4 - \beta)u(\beta - 2)).$$

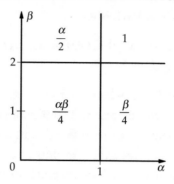

FIGURE 4.9: CDF representation for Example 4.1.7.

(b) The marginal CDFs are found as

$$F_x(\alpha) = F_{x,y}(\alpha, \infty) = \frac{\alpha}{2}u(\alpha) + \left(1 - \frac{\alpha}{2}\right)u(\alpha - 1)$$

and

$$F_y(\beta) = F_{x,y}(\infty, \beta) = \frac{\beta}{4}u(\beta) + \left(1 - \frac{\beta}{4}\right)u(\beta - 2).$$

Since $F_{x,y}(0.5, 0.5) = 1/16 \neq 1/32 = F_x(0.5)F_y(0.5)$, we conclude that x and y are not independent. (c) Differentiating, we find

$$f_x(\alpha) = 0.5(u(\alpha) - u(\alpha - 1)) + 0.5\delta(\alpha - 1)$$

and

$$f_y(\beta) = 0.25(u(\beta) - u(\beta - 2)) + 0.5\delta(\beta - 2).$$

Partial differentiation of $F_{x,y}(\alpha, \beta)$ with respect to α and β yields

$$f_{x,y}(\alpha, \beta) = 0.25(u(\alpha) - u(\alpha - 1))(u(\beta) - u(\beta - 2)) + 0.5\delta(\alpha - 1)\delta(\beta - 2).$$

This differentiation result can of course be obtained using the product rule and using $u^{(1)}(\alpha) = \delta(\alpha)$. An easier way is to use the two-dimensional representation of Fig. 4.9. Inside any of the indicated regions, the CDF is easily differentiated. If there is a jump along the boundary, then there is a Dirac delta function in the variable which changes to move across the boundary. An examination of Fig. 4.9 reveals a jump of 0.5 along $\beta = 2$, $1 \leq \alpha$. Another jump of height 0.5 occurs along $\alpha = 1$, $2 \leq \beta$. Since errors are always easily made, it is always worthwhile to check the result by integrating the resulting PDF to ensure the total volume under the PDF is one.

(d) The given integral is

$$I = \int_{-\infty}^{\infty} F_x(\beta) f_y(\beta) \, d\beta.$$

Substituting, we find

$$I = \int_0^1 \frac{\beta}{2} \left(\frac{1}{4}(u(\beta) - u(\beta - 2)) + \frac{1}{2}\delta(\beta - 2) \right) d\beta$$

$$+ \int_1^{\infty} \left(\frac{1}{4}(u(\beta) - u(\beta - 2)) + \frac{1}{2}\delta(\beta - 2) \right) d\beta$$

$$= \int_0^1 \frac{\beta}{8} \, d\beta + \int_1^2 \frac{1}{4} \, d\beta + \frac{1}{2} = \frac{13}{16}.$$

(e) We have

$$P(x \leq y) = \int_{-\infty}^{\infty} \int_{-\infty}^{\beta} f_{x,y}(\alpha, \beta) \, d\alpha \, d\beta.$$

Substituting,

$$P(x \leq y) = \int_0^1 \int_0^{\beta} \frac{1}{4} \, d\alpha \, d\beta + \int_1^2 \int_0^1 \frac{1}{4} \, d\alpha \, d\beta + \frac{1}{2} = \frac{7}{8}.$$

As an alternative, $P(x \leq y) = 1 - P(x > y)$, with

$$P(x > y) = \int_0^1 \int_0^{\alpha} \frac{1}{4} \, d\beta d\alpha = \frac{1}{8}.$$

■

Drill Problem 4.1.1. *Consider the experiment of tossing a fair coin three times. Let the random variable x denote the total number of heads and the random variable y denote the difference between the number of heads and tails resulting from the experiment. Determine: (a) $p_{x,y}(3, 3)$, (b) $p_{x,y}(1, -1)$, (c) $p_{x,y}(2, 1)$, (d) $p_{x,y}(0, -3)$, (e) $F_{x,y}(0, 0)$, (f) $F_{x,y}(1, 8)$, (g) $F_{x,y}(2, 1)$, and (h) $F_{x,y}(3, 3)$.*

Answers: 1/8, 3/8, 1/8, 3/8, 1, 1/2, 7/8, 1/8.

Drill Problem 4.1.2. *The RVs x and y have joint PMF specified in the table below.*

α	β	$p_{x,y}(\alpha, \beta)$
0	1	1/8
1	1	1/8
1	2	2/8
1	3	1/8
2	2	1/8
2	3	1/8
3	3	1/8

Determine: (a) $p_x(1)$, (b) $p_y(2)$, (c) $p_x(2)$, (d) $p_x(3)$.

Answers: 1/8, 1/4, 1/2, 3/8.

Drill Problem 4.1.3. *Consider the experiment of tossing a fair tetrahedral die (with faces labeled 0,1,2,3) twice. Let x be a RV equaling the sum of the numbers tossed, and let y be a RV equaling the absolute value of the difference of the numbers tossed. Find: (a) $F_y(0)$, (b) $F_y(2)$, (c) $p_y(2)$, (d) $p_y(3)$.*

Answers: 1/4, 14/16, 4/16, 2/16.

Drill Problem 4.1.4. *The joint PDF for the RVs x and y is*

$$f_{x,y}(\alpha, \beta) = \begin{cases} 2\dfrac{\beta}{\alpha}, & 0 < \beta \leq \sqrt{\alpha} < 1 \\ 0, & \text{elsewhere.} \end{cases}$$

Find: (a) $f_x(0.25)$, (b) $f_y(0.25)$, (c) whether or not x and y are independent random variables.

Answers: 1, ln (4), no.

Drill Problem 4.1.5. *With the joint PDF of random variables x and y given by*

$$f_{x,y}(\alpha, \beta) = \begin{cases} a\alpha^2\beta, & 0 \leq \alpha \leq 3, 0 \leq \beta \leq 1 \\ 0, & \text{otherwise,} \end{cases}$$

where a is a constant, determine: (a) a, (b) $P(0 \leq x \leq 1, 0 \leq y \leq 1/2)$, (c) $P(xy \leq 1)$, (d) $P(x + y \leq 1)$.

Answers: 1/108, 7/27, 2/9, 1/270.

Drill Problem 4.1.6. *With the joint PDF of random variables x and y given by*

$$f_{x,y}(\alpha, \beta) = \begin{cases} a\alpha\beta(1 - \alpha), & 0 \leq \alpha \leq 1 - \beta \leq 1 \\ 0, & \text{otherwise,} \end{cases}$$

where a is a constant, determine: (a) a, (b) $f_x(0.5)$, (c) $F_x(0.5)$, (d) $F_y(0.25)$.

Answers: 13/16, 49/256, 5/4, 40.

4.2   BIVARIATE RIEMANN-STIELTJES INTEGRAL

The Riemann-Stieltjes integral provides a unified framework for treating continuous, discrete, and mixed RVs—all with one kind of integration. An important alternative is to use a standard Riemann integral for continuous RVs, a summation for discrete RVs, and a Riemann integral with an integrand containing Dirac delta functions for mixed RVs. In the following, we assume that F is the joint CDF for the RVs x and y, that $a_1 < b_1$, and that $a_2 < b_2$.

We begin with a brief review of the standard Riemann integral. Let

$$a_1 = \alpha_0 < \alpha_1 < \alpha_2 < \cdots < \alpha_n = b_1,$$

$$a_2 = \beta_0 < \beta_1 < \beta_2 < \cdots < \beta_m = b_2,$$

$$\alpha_{i-1} \leq \xi_i \leq \alpha_i, \qquad i = 1, 2, \ldots, n,$$

$$\beta_{j-1} \leq \psi_j \leq \beta_j, \qquad j = 1, 2, \ldots, m,$$

$$\Delta_{1,n} = \max_{1 \leq i \leq n} \{\alpha_i - \alpha_{i-1}\},$$

and

$$\Delta_{2,m} = \max_{1 \leq j \leq m} \{\beta_j - \beta_{j-1}\}.$$

The **Riemann integral**

$$\int_{a_2}^{b_2} \int_{a_1}^{b_1} h(\alpha, \beta) \, d\alpha \, d\beta$$

is defined by

$$\lim_{\Delta_{2,m} \to 0} \lim_{\Delta_{1,n} \to 0} \sum_{j=1}^{m} \sum_{i=1}^{n} h(\xi_i, \psi_j)(\alpha_i - \alpha_{i-1})(\beta_j - \beta_{j-1}),$$

provided the limits exist and are independent of the choice of $\{\xi_i\}$ and $\{\psi_j\}$. Note that $n \to \infty$ and $m \to \infty$ as $\Delta_{1,n} \to 0$ and $\Delta_{2,m} \to 0$. The summation above is called a Riemann sum. We remind the reader that this is the "usual" integral of calculus and has the interpretation as the volume under the surface $h(\alpha, \beta)$ over the region $a_1 < \alpha < b_1, a_2 < \beta < b_2$.

With the same notation as above, the **Riemann-Stieltjes integral**

$$\int\limits_{a_2}^{b_2} \int\limits_{a_1}^{b_1} g(\alpha, \beta) \, dF(\alpha, \beta) h(\alpha, \beta) \, d\alpha \, d\beta$$

is defined by

$$\lim_{\Delta_{2,m} \to 0} \lim_{\Delta_{1,n} \to 0} \sum_{j=1}^{m} \sum_{i=1}^{n} g(\xi_i, \psi_j) \Delta_2(\beta_j, \beta_{j-1}) \Delta_1(\alpha_i - \alpha_{i-1}) F(\alpha_i, \beta_j),$$

provided the limits exist and are independent of the choice of $\{\xi_i\}$ and $\{\psi_j\}$.

Applying the above definition with $g(\alpha, \beta) \equiv 1$, we obtain

$$\int\limits_{a_2}^{b_2} \int\limits_{a_1}^{b_1} dF(\alpha, \beta) = \lim_{\Delta_{2,m} \to 0} \sum_{j=1}^{m} \Delta_2(\beta_j - \beta_{j-1})(F(b_1, \beta_j) - F(\alpha_1, \beta_j))$$

$$= F(b_1, b_2) - F(a_1, b_2) - (F(b_1, a_2) - F(a_1, a_2))$$

$$= P(a_1 < x \le b_1, a_2 < y \le b_2).$$

Suppose F is discrete with jumps at $(\alpha, \beta) \in \{(\alpha_i, \beta_i) : i = 0, 1, \ldots N\}$ satisfying

$$a_1 = \alpha_0 < \alpha_1 < \cdots < \alpha_N \le b_1$$

and

$$a_2 = \beta_0 < \beta_1 < \cdots < \beta_N \le b_2.$$

Then, provided that g and F have no common points of discontinuity, it is easily shown that

$$\int\limits_{a_2}^{b_2} \int\limits_{a_1}^{b_1} g(\alpha, \beta) \, dF(\alpha, \beta) = \sum_{i=1}^{N} g(\alpha_i, \beta_i) p(\alpha_i, \beta_i), \qquad (4.33)$$

where

$$p(\alpha, \beta) = F(\alpha, \beta) - F(\alpha^-, \beta) - (F(\alpha, \beta^-) - F(\alpha^-, \beta^-)). \qquad (4.34)$$

Note that a jump in F at (a_1, a_2) is not included in the sum whereas a jump at (b_1, b_2) is included. Suppose F is absolutely continuous with

$$f(\alpha, \beta) = \frac{\partial^2 F(\alpha, \beta)}{\partial \beta \, \partial \alpha}. \tag{4.35}$$

Then

$$\int_{a_2}^{b_2} \int_{a_1}^{b_1} g(\alpha, \beta) \, dF(\alpha, \beta) = \int_{a_2}^{b_2} \int_{a_1}^{b_1} g(\alpha, \beta) f(\alpha, \beta) \, d\alpha \, d\beta. \tag{4.36}$$

Hence, the Riemann-Stieltjes integral reduces to the usual Riemann integral in this case. Formally, we may write

$$\begin{aligned}
dF(\alpha, \beta) &= \lim_{b_2 \to 0} \lim_{b_1 \to 0} \frac{\Delta_2(h_2) \Delta_1(h_1) F(\alpha, \beta)}{h_1 h_2} \, d\alpha \, d\beta \\
&= \frac{\partial^2 F(\alpha, \beta)}{\partial \beta \, \partial \alpha} \, d\alpha \, d\beta,
\end{aligned} \tag{4.37}$$

provided the indicated limits exist. The major advantage of the Riemann-Stieltjes integral is to enable one to evaluate the integral in many cases where the above limits do not exist. For example, with

$$F(\alpha, \beta) = u(\alpha - 1) u(\beta - 2)$$

we may write

$$dF(\alpha, \beta) = du(\alpha - 1) \, du(\beta - 2).$$

The trick to evaluating the Riemann-Stieltjes integral involves finding a suitable approximation for

$$\Delta_2(h_2) \Delta_1(h_1) F(\alpha, \beta)$$

which is valid for small h_1 and small h_2.

Example 4.2.1. *The RVs x and y have joint CDF*

$$\begin{aligned}
F_{x,y}(\alpha, \beta) &= \frac{1}{2}(1 - e^{-2\alpha})(1 - e^{-3\beta}) u(\alpha) u(\beta) \\
&\quad + \frac{1}{8} u(\alpha) u(\beta + 2) + \frac{3}{8} u(\alpha - 1) u(\beta - 4).
\end{aligned}$$

Find $P(x > y)$.

Solution. For this example, we obtain

$$dF_{x,y}(\alpha, \beta) = 3e^{-2\alpha}e^{-3\beta}u(\alpha)u(\beta)\,d\alpha\,d\beta$$
$$+\frac{1}{8}\,du(\alpha)\,du(\beta+2) + \frac{3}{8}\,du(\alpha-1)\,du(\beta-4).$$

Consequently,

$$P(x > y) = \int_{-\infty}^{\infty}\int_{\beta}^{\infty} dF_{x,y}(\alpha, \beta)$$

$$= \int_{0}^{\infty}\int_{\beta}^{\infty} 3e^{-2\alpha}e^{-3\beta}\,d\alpha\,d\beta + \frac{1}{8}$$

$$= \int_{0}^{\infty} 3\frac{0 - e^{-2\beta}}{-2}e^{-3\beta}\,d\beta + \frac{1}{8}$$

$$= \frac{3}{2}\frac{0-1}{-5} + \frac{1}{8} = \frac{9}{40}.$$

∎

Example 4.2.2. *The RVs x and y have joint CDF with two-dimensional representation shown in Fig. 4.10. The CDF $F_{x,y}(\alpha, \beta) = 0$ for $\alpha < 0$ or $\beta < 0$. (a) Find a suitable expression for $dF_{x,y}(\alpha, \beta)$. Verify by computing $F_{x,y}$. (b) Find $P(x = 2y)$. (c) Evaluate*

$$I = \int_{-\infty}^{\infty}\int_{-\infty}^{\infty} \alpha\beta\,dF_{x,y}(\alpha, \beta).$$

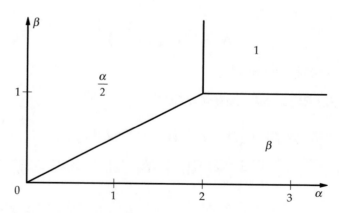

FIGURE 4.10: Cumulative distribution function for Example 4.2.2.

Solution.   (a) Careful inspection of Fig. 4.10 reveals that the CDF is continuous (everywhere) and that $dF_{x,y}(\alpha, \beta) = 0$ everywhere except along $0 < \alpha = 2\beta < 2$. We conclude that

$$dF_{x,y}(\alpha, \beta) = dF_x(\alpha)\, du\left(\beta - \frac{\alpha}{2}\right) = dF_y(\beta)\, du(\alpha - 2\beta).$$

To support this conclusion, we find

$$\int\limits_{-\infty}^{\beta}\int\limits_{-\infty}^{\alpha} dF_x(\alpha')\, du\left(\beta' - \frac{\alpha'}{2}\right) = \int\limits_{-\infty}^{\alpha}\left(\int\limits_{-\infty}^{\beta} du\left(\beta' - \frac{\alpha'}{2}\right)\right) dF_x(\alpha')$$

$$= \int\limits_{-\infty}^{\alpha} u\left(\beta - \frac{\alpha'}{2}\right) dF_x(\alpha')$$

$$= F_x(\min(\{\alpha, 2\beta\})) = F_{x,y}(\alpha, \beta).$$

Similarly,

$$\int\limits_{-\infty}^{\beta}\left(\int\limits_{-\infty}^{\alpha} du(\alpha' - 2\beta')\right) dF_y(\beta') = \int\limits_{-\infty}^{\beta} u(\alpha - 2\beta')\, dF_y(\beta')$$

$$= F_y(\min(\{0.5\alpha, \beta\})) = F_{x,y}(\alpha, \beta).$$

(b) From part (a) we conclude that $P(x = 2y) = 1$.

(c) Using results of part (a),

$$I = \int\limits_{-\infty}^{\infty} \frac{\alpha^2}{2}\, dF_x(\alpha) = \int\limits_{0}^{2} \frac{\alpha^2}{4}\, d\alpha = \frac{8 - 0}{12} = \frac{2}{3}.$$

We note that

$$I = E(xy) = E(2y^2) = 2\int\limits_{0}^{1} \beta^2\, d\beta = \frac{2}{3}.$$

■

4.3   EXPECTATION

Expectation involving jointly distributed RVs is quite similar to the univariate case. The basic difference is that two-dimensional integrals are required.

4.3.1   Moments

Definition 4.3.1.   *The **expected value** of $g(x, y)$ is defined by*

$$E(g(x, y)) = \int_{-\infty}^{\infty} \int_{-\infty}^{\infty} g(\alpha, \beta) \, dF_{x,y}(\alpha, \beta), \qquad (4.38)$$

*provided the integral exists. The **mean** of the bivariate RV $\mathbf{z} = (x, y)$ is defined by*

$$\eta_z = (\eta_x, \eta_y). \qquad (4.39)$$

*The **covariance** of the RVs x and y is defined by*

$$\sigma_{x,y} = E((x - \eta_x)(y - \eta_y)). \qquad (4.40)$$

*The **correlation coefficient** of the RVs x and y is defined by*

$$\rho_{x,y} = \frac{\sigma_{x,y}}{\sigma_x \sigma_y}. \qquad (4.41)$$

*The **joint (m,n)th moment** of x and y is*

$$m_{m,n} = E(x^m y^n), \qquad (4.42)$$

*and the **joint (m,n)th central moment** of x and y is*

$$\mu_{m,n} = E((x - \eta_x)^m (y - \eta_y)^n). \qquad (4.43)$$

Definition 4.3.2.   *The joint RVs x and y are **uncorrelated** if*

$$E(xy) = E(x)E(y), \qquad (4.44)$$

*and **orthogonal** if*

$$E(xy) = 0. \qquad (4.45)$$

Theorem 4.3.1.   *If the RVs x and y are independent, then*

$$E(g(x)h(y)) = E(g(x))E(h(y)). \qquad (4.46)$$

Proof.   Since x and y are independent, we have $F_{x,y}(\alpha, \beta) = F_x(\alpha)F_y(\beta)$ so that $dF_{x,y}(\alpha, \beta) = dF_x(\alpha) \, dF_y(\beta)$. Consequently,

$$E(g(x)h(y)) = \int_{-\infty}^{\infty} \int_{-\infty}^{\infty} g(\alpha)h(\beta) \, dF_x(\alpha) \, dF_y(\beta) = E(g(x))E(h(y)).$$

■

Theorem 4.3.2.   *The RVs x and y are uncorrelated iff $\sigma_{x,y} = 0$. If x and y are uncorrelated and $\eta_x = 0$ and/or $\eta_y = 0$, then x and y are orthogonal.*

Note that if x and y are independent, then x and y are uncorrelated; the converse is not true, in general.

Example 4.3.1.   *RV x has PDF*

$$f_x(\alpha) = \frac{1}{4}(u(\alpha) - u(\alpha - 4))$$

and RV $y = ax + b$, where a and b are real constants with $a \neq 0$. Find: (a) $E(xy)$, (b) σ_x^2, (c) σ_y^2, (d) $\rho_{x,y}$.

Solution.   (a) We have

$$E(x) = \frac{1}{4} \int\limits_0^4 \alpha \, d\alpha = \frac{16}{8} = 2,$$

$$E(x^2) = \frac{1}{4} \int\limits_0^4 \alpha^2 \, d\alpha = \frac{64}{12} = \frac{16}{3},$$

so that

$$E(xy) = E(ax^2 + bx) = \frac{16}{3}a + 2b.$$

Note that x and y are orthogonal if

$$\frac{16}{3}a + 2b = 0.$$

(b) $\sigma_x^2 = E(x^2) - E^2(x) = \frac{16}{3} - 4 = \frac{4}{3}$.

(c) $\sigma_y^2 = E((ax + b - 2a - b)^2) = a^2\sigma_x^2$.

(d) Noting that $\sigma_{x,y} = a\sigma_x^2$ we find

$$\rho_{x,y} = \frac{\sigma_{x,y}}{\sigma_x\sigma_y} = \frac{a}{|a|}.$$

Note that $\rho_{x,y} = -1$ if $a < 0$ and $\rho = 1$ if $a > 0$. The correlation coefficient provides information about how x and y are related to each other. Clearly, if $x = y$ then $\rho_{x,y} = 1$. This example also shows that if there is a linear relationship between x and y then $\rho = \pm 1$.   ∎

Example 4.3.2. *RVs x and y are uncorrelated, and RV z = x + y. Find: (a) $E(z^2)$, (b) σ_z^2.*

Solution. (a) Using the properties of expectation,

$$E(z^2) = E(x^2 + 2xy + y^2) = E(x^2) + 2\eta_x\eta_y + E(y^2).$$

(b) With $z = x + y$,

$$\sigma_z^2 = E((x - \eta_x + y - \eta_y)^2) = \sigma_x^2 + 2\sigma_{x,y} + \sigma_y^2.$$

Since x and y are uncorrelated we have $\sigma_{x,y} = 0$ so that $\sigma_z^2 = \sigma_x^2 + \sigma_y^2$; i.e., the variance of the sum of uncorrelated RVs is the sum of the individual variances.   ∎

Example 4.3.3. *Random variables x and y have the joint PMF shown in Fig. 4.5. Find $E(x + y)$, $\sigma_{x,y}$, and $\rho_{x,y}$.*

Solution. We have

$$E(x + y) = \sum_{(\alpha,\beta)} (\alpha + \beta) p_{x,y}(\alpha, \beta).$$

Substituting,

$$E(x + y) = 0 \cdot \frac{1}{4} - 1 \cdot \frac{1}{8} + 0 \cdot \frac{1}{8} + 2 \cdot \frac{1}{8} + 3 \cdot \frac{1}{8} + 3 \cdot \frac{1}{8} + 6 \cdot \frac{1}{8} = \frac{13}{8}.$$

In order to find $\sigma_{x,y}$, we first find η_x and η_y:

$$\eta_x = -1 \cdot \frac{1}{4} + 0 \cdot \frac{1}{8} + 1 \cdot \frac{3}{8} + 2 \cdot \frac{1}{8} + 3 \cdot \frac{1}{8} = \frac{3}{4},$$

and

$$\eta_y = -1 \cdot \frac{1}{4} + 0 \cdot \frac{1}{8} + 1 \cdot \frac{3}{8} + 3 \cdot \frac{2}{8} = \frac{7}{8}.$$

Then

$$\sigma_{x,y} = E(xy) - \eta_x\eta_y = -1 \cdot \frac{1}{8} - 1 \cdot \frac{1}{4} + 1 \cdot \frac{1}{8} + 2 \cdot \frac{1}{8} + 9 \cdot \frac{1}{8} - \frac{3}{4} \cdot \frac{7}{8} = \frac{15}{32}.$$

We find

$$E(x^2) = 1 \cdot \frac{1}{4} + 0 \cdot \frac{1}{8} + 1 \cdot \frac{3}{8} + 4 \cdot \frac{1}{8} + 9 \cdot \frac{1}{8} = \frac{9}{4},$$

and

$$E(y^2) = 1 \cdot \frac{1}{4} + 0 \cdot \frac{1}{8} + 1 \cdot \frac{3}{8} + 9 \cdot \frac{2}{8} = \frac{23}{8},$$

so that $\sigma_x = \sqrt{27/16} = 1.299$ and $\sigma_y = \sqrt{135/64} = 1.4524$. Finally,

$$\rho_{x,y} = \frac{\sigma_{x,y}}{\sigma_x \sigma_y} = 0.2485.$$ ∎

Example 4.3.4. *Random variables x and y have joint PDF*

$$f_{x,y}(\alpha, \beta) = \begin{cases} 1.5(\alpha^2 + \beta^2), & 0 < \alpha < 1, 0 < \beta < 1, \\ 0, & \text{elsewhere.} \end{cases}$$

Find $\sigma_{x,y}$.

Solution. Since $\sigma_{x,y} = E(xy) - \eta_x \eta_y$, we find

$$E(x) = \int_0^1 \int_0^1 \alpha 1.5(\alpha^2 + \beta^2) \, d\alpha \, d\beta = \frac{5}{8}.$$

Due to the symmetry of the PDF, we find that $E(y) = E(x) = 5/8$. Next

$$E(xy) = \int_0^1 \int_0^1 \alpha \beta 1.5(\alpha^2 + \beta^2) \, d\alpha \, d\beta = \frac{3}{8}.$$ ∎

Finally, $\sigma_{x,y} = -3/192$.

The moment generating function is easily extended to two dimensions.

Definition 4.3.3. *The **joint moment generating function** for the RVs x and y is defined by*

$$M_{x,y}(\lambda_1, \lambda_2) = E(e^{\lambda_1 x + \lambda_2 y}),$$ *(4.47)*

where λ_1 and λ_2 are real variables.

Theorem 4.3.3. *Define*

$$M_{x,y}^{(m,n)}(\lambda_1, \lambda_2) = \frac{\partial^{m+n} M_{x,y}(\lambda_1, \lambda_2)}{\partial \lambda_1^m \, \partial \lambda_2^n}.$$ *(4.48)*

The (m,n)th joint moment for x and y is given by

$$E(x^m y^n) = M_{x,y}^{(m,n)}(0, 0).$$ *(4.49)*

Example 4.3.5. *The joint PDF for random variables x and y is given by*

$$f_{x,y}(\alpha, \beta) = \begin{cases} a \, e^{-|\alpha+\beta|}, & 0 < \beta < 1 \\ 0, & \text{otherwise.} \end{cases}$$

Determine: (a) $M_{x,y}$; (b) a; (c) $M_x(\lambda)$ and $M_y(\lambda)$; (d) $E(x), E(y)$, and $E(xy)$.

Solution. (a) Using the definition of moment generating function,

$$M_{x,y}(\lambda_1, \lambda_2) = a \int\limits_{0}^{1} \left(\int\limits_{-\infty}^{-\beta} e^{\alpha(\lambda_1+1)+\beta} \, d\alpha + \int\limits_{-\beta}^{\infty} e^{\alpha(\lambda_1-1)-\beta} \, d\alpha \right) e^{\lambda_2 \beta} \, d\beta.$$

The first inner integral converges for $-1 < \lambda_1$, the second converges for $\lambda_1 < 1$. Straightforward integration yields $(-1 < \lambda_1 < 1)$

$$M_{x,y}(\lambda_1, \lambda_2) = 2ag(\lambda_1 - \lambda_2)h(\lambda_1),$$

where $g(\lambda) = (1 - e^{-\lambda})/\lambda$ and $h(\lambda) = 1/(1 - \lambda^2)$.

(b) Since $M_{x,y}(0, 0) = E(e^0) = 1$, applying L'Hôpital's Rule, we find $M_{x,y}(0, 0) = 2a$, so that $a = 0.5$.

(c) We obtain $M_x(\lambda) = M_{x,y}(\lambda, 0) = g(\lambda)h(\lambda)$. Similarly, $M_y(\lambda) = M_{x,y}(0, \lambda) = g(-\lambda)$.

(d) Differentiating, we have

$$M_x^{(1)}(\lambda) = g^{(1)}(\lambda)h(\lambda) + g(\lambda)h^{(1)}(\lambda),$$

$$M_y^{(1)}(\lambda) = -g^{(1)}(-\lambda),$$

and

$$M_{x,y}^{(1,1)}(\lambda_1, \lambda_2) = -g^{(2)}(\lambda_1 - \lambda_2)h(\lambda_1) - g^{(1)}(\lambda_1 - \lambda_2)h^{(1)}(\lambda_1).$$

Noting that

$$g(\lambda) = 1 - \frac{\lambda}{2} + \frac{\lambda^2}{6} - \frac{\lambda^3}{24} + \cdots,$$

we find easily that $g(0) = 1$, $g^{(1)}(0) = -0.5$, and $g^{(2)}(0) = 1/3$. Since $h^{(1)}(0) = 0$, we obtain $E(x) = -0.5$, $E(y) = 0.5$, and $E(xy) = -1/3$. ∎

4.3.2   Inequalities

Theorem 4.3.4.   (Hölder Inequality) *Let p and q be real constants with $p > 1, q > 1$, and*

$$\frac{1}{p} + \frac{1}{q} = 1. \tag{4.50}$$

If x and y are RVs with $a = E^{1/p}(|x|^p) < \infty$ and $b = E^{1/q}(|y|^q) < \infty$ then

$$E(|xy|) \le E^{1/p}(|x|^p)E^{1/q}(|y|^q). \tag{4.51}$$

Proof. If either $a = 0$ or $b = 0$ then $P(xy = 0) = 1$ so that $E(|xy|) = 0$; hence, assume $a > 0$ and $b > 0$. Let

$$g(\alpha) = \frac{\alpha^p}{p} + \frac{\beta^q}{q} - \alpha\beta,$$

for $\alpha \geq 0$, $\beta > 0$. We have $g(0) > 0$, $g(\infty) = \infty$, $g^{(1)}(\alpha) = \alpha^{p-1} - \beta$, and $g^{(2)}(\alpha_0) = (p-1)\alpha_0^{p-2} > 0$, where α_0 satisfies $g^{(1)}(\alpha_0) = 0$. Thus, $g(\alpha) \geq g(\alpha_0)$, and $\alpha_0 = \beta^{1/(p-1)} = \beta^{q/p}$. Consequently,

$$\frac{\alpha^p}{p} + \frac{\beta^q}{q} - \alpha\beta \geq \frac{\alpha_0^p}{p} + \frac{\beta^q}{q} - \alpha_0\beta = 0.$$

The desired result follows by letting $\alpha = |x|/a$ and $\beta = |y|/b$.   ∎

Corollary 4.3.1. (Schwarz Inequality)

$$E^2(|xy|) \leq E(|x|^2)E(|y|^2). \tag{4.52}$$

If $y = ax$, for some constant a, then

$$E^2(|xy|) = |a|^2 E^2(|x|^2) = E(|x|^2)E(|y|^2).$$

Applying the Schwarz Inequality, we find that the covariance between x and y satisfies

$$\sigma_{x,y}^2 = E^2((x - \eta_x)(y - \eta_y)) \leq \sigma_x^2 \sigma_y^2.$$

Hence, the correlation coefficient satisfies

$$|\rho_{x,y}| \leq 1. \tag{4.53}$$

If there is a linear relationship between the RVs x and y, then $|\rho_{x,y}| = 1$, as shown in Example 4.3.1.

Theorem 4.3.5. (Minkowski Inequality) *Let p be a real constant with $p \geq 1$. If x and y are RVs with $E(|x|^p) < \infty$ and $E(|y|^p) < \infty$ then*

$$E^{1/p}(|x + y|^p) \leq E^{1/p}(|x|^p) + E^{1/p}(|y|^p). \tag{4.54}$$

Proof. From the triangle inequality $(|x + y| \leq |x| + |y|)$,

$$E(|x + y|^p) = E(|x + y||x + y|^{p-1})$$
$$\leq E(|x||x + y|^{p-1}) + E(|y||x + y|^{p-1}),$$

which yields the desired result if $p = 1$. For $p > 1$, let $q = p/(p - 1)$ and apply the Hölder Inequality to obtain

$$E(|x + y|^p) \leq E^{1/p}(|x|^p)E^{1/q}(|x + y|^p) + E^{1/p}(|y|^p)E^{1/q}(|x + y|^p),$$

from which the desired result follows. ∎

Theorem 4.3.6. *With $\alpha_k = E^{1/k}(|x|^k)$ we have $\alpha_{k+1} \geq \alpha_k$ for $k = 1, 2, \ldots$.*

Proof. Let $\beta_i = E(|x|^i)$. From the Schwarz inequality,

$$\beta_i^2 = E^2(|x|^{(i-1)/2}|x|^{(i+1)/2}) \leq E(|x|^{i-1})E(|x|^{i+1}) = \beta_{i-1}\beta_{i+1}.$$

Raising to the ith power and taking the product (noting that $\beta_0 = 1$)

$$\prod_{i=1}^{k} \beta_i^{2i} \leq \prod_{i=1}^{k} \beta_{i-1}^i \beta_{i+1}^i = \prod_{i=0}^{k-1} \beta_i^{i+1} \prod_{j=2}^{k+1} \beta_j^{j-1} = \beta_k^{k-1} \beta_{k+1}^k \prod_{i=1}^{k-1} \beta_i^{2i}.$$

Simplifying, we obtain $\beta_k^{k+1} \leq \beta_{k+1}^k$; the desired inequality follows by raising to the $1/(k(k + 1))$ power. ∎

4.3.3 Joint Characteristic Function

Definition 4.3.4. *The **joint characteristic** function for the RVs x and y is defined by*

$$\phi_{x,y}(t_1, t_2) = E(e^{jxt_1 + jyt_2}), \tag{4.55}$$

where t_1 and t_2 are real variables, and $j^2 = -1$.

Note that the marginal characteristic functions ϕ_x and ϕ_y are easily obtained from the joint characteristic function as $\phi_x(t) = \phi_{x,y}(t, 0)$ and $\phi_y(t) = \phi_{x,y}(0, t)$.

Theorem 4.3.7. *The joint RVs x and y are independent iff*

$$\phi_{x,y}(t_1, t_2) = \phi_x(t_1)\phi_y(t_2) \tag{4.56}$$

for all real t_1 and t_2.

Theorem 4.3.8. *If x and y are independent RVs, then*

$$\phi_{x+y}(t) = \phi_x(t)\phi_y(t). \tag{4.57}$$

Theorem 4.3.9. *The joint (m,n)th moment of the RVs x and y can be obtained from the joint characteristic function as*

$$E(x^m y^n) = (-j)^{m+n} \phi_{x,y}^{(m,n)}(0, 0). \tag{4.58}$$

The joint characteristic function $\phi_{x,y}$ contains all of the information about the joint CDF $F_{x,y}$; the joint CDF itself can be obtained from the joint characteristic function.

Theorem 4.3.10. *If the joint CDF $F_{x,y}$ is continuous at (a_1, a_2) and at (b_1, b_2), with $a_1 < b_1$ and $a_2 < b_2$, then*

$$P(a_1 < x \leq b_1, a_2 < y \leq b_2) = \lim_{T \to \infty} \int_{-T}^{T} \int_{-T}^{T} \frac{e^{-ja_1 t_1} - e^{-jb_1 t_1}}{j2\pi t_1} \frac{e^{-ja_2 t_2} - e^{-jb_2 t_2}}{j2\pi t_2} \phi_{x,y}(t_1, t_2) \, dt_1 \, dt_2.$$

$$(4.59)$$

Proof. The proof is a straightforward extension of the corresponding one-dimensional result. ∎

Corollary 4.3.2. *If x and y are jointly continuous RVs with $\phi_{x,y}$, then*

$$f_{x,y}(\alpha, \beta) = \lim_{T \to \infty} \frac{1}{(2\pi)^2} \int_{-T}^{T} \int_{-T}^{T} e^{-j\alpha t_1 - j\beta t_2} \phi_{x,y}(t_1, t_2) \, dt_1 \, dt_2. \qquad (4.60)$$

The above corollary establishes that the joint PDF is $1/(2\pi)^2$ times the two-dimensional Fourier transform of the joint characteristic function.

Drill Problem 4.3.1. *The joint PDF for RVs x and y is*

$$f_{x,y}(\alpha, \beta) = \begin{cases} \dfrac{2}{9}\alpha^2 \beta, & 0 < \alpha < 3, 0 < \beta < 1 \\ 0, & \text{otherwise.} \end{cases}$$

Find $\sigma_{x,y}$.

Answer: 0.

Drill Problem 4.3.2. *Suppose the RVs x and y have the joint PMF shown in Fig. 4.11. Determine:* *(a) $E(x)$, (b) $E(y)$, (c) $E(x + y)$, and (d) $\sigma_{x,y}$.*

Answers: 0.54, 1.6, 3.2, 1.6.

Drill Problem 4.3.3. *Suppose $\eta_x = 5$, $\eta_y = 3$, $\sigma_{x,y} = 18$, $\sigma_x = 3$, and $\sigma_y = 6$. Find: (a) $E(x^2)$, (b) $E(xy)$, (c) σ_{3x}^2, and (d) σ_{x+y}^2.*

Answers: 81, 81, 33, 34.

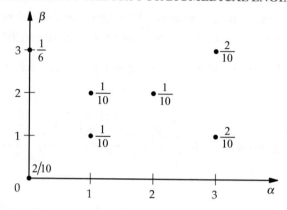

FIGURE 4.11: PMF for Drill Problem 4.3.2.

4.4   CONVOLUTION

The convolution operation arises in many applications. Convolution describes the basic input/output relationship for a linear, time-invariant system, as well as the distribution function for the sum of two independent RVs.

Theorem 4.4.1. *If x and y are independent RVs and $z = x + y$ then*

$$F_z(\gamma) = \int_{-\infty}^{\infty} F_x(\gamma - \beta)dF_y(\beta) = \int_{-\infty}^{\infty} F_y(\gamma - \alpha)dF_x(\alpha). \qquad (4.61)$$

*The above integral operation on the functions F_x and F_y is called a **convolution**.*

Proof. By definition,

$$F_z(\gamma) = P(z \leq \gamma) = \int_{\alpha + \beta \leq \gamma} dF_{x,y}(\alpha, \beta).$$

Since x and y are independent, we have

$$F_z(\gamma) = \int_{-\infty}^{\infty} \int_{-\infty}^{\gamma - \beta} dF_x(\alpha) \, dF_y(\beta) = \int_{-\infty}^{\infty} F_x(\gamma - \beta) \, dF_y(\beta).$$

Interchanging the order of integration,

$$F_z(\gamma) = \int_{-\infty}^{\infty} \int_{-\infty}^{\gamma - \alpha} dF_y(\beta) \, dF_x(\alpha) = \int_{-\infty}^{\infty} F_y(\gamma - \alpha) \, dF_x(\alpha).$$

∎

Corollary 4.4.1. *Let x and y be independent RVs and let z = x + y.*

(i) If x is a continuous RV then z is a continuous RV with PDF

$$f_z(\gamma) = \int_{-\infty}^{\infty} f_x(\gamma - \beta) \, dF_y(\beta). \qquad (4.62)$$

(ii) If y is a continuous RV then z is a continuous RV with PDF

$$f_z(\gamma) = \int_{-\infty}^{\infty} f_y(\gamma - \alpha) \, dF_x(\alpha). \qquad (4.63)$$

(iii) If x and y are jointly continuous RVs then z is a continuous RV with PDF

$$f_z(\gamma) = \int_{-\infty}^{\infty} f_x(\gamma - \beta) f_y(\beta) \, d\beta = \int_{-\infty}^{\infty} f_y(\gamma - \alpha) f_x(\alpha) \, d\alpha. \qquad (4.64)$$

(iv) If x and y are both discrete RVs then z is a discrete RV with PMF

$$p_z(\gamma) = \sum_{\beta} p_x(\gamma - \beta) p_y(\beta) = \sum_{\alpha} p_y(\gamma - \alpha) p_x(\alpha). \qquad (4.65)$$

All of these operations are called convolutions.

Example 4.4.1. *Random variables x and y are independent with $f_x(\alpha) = 0.5(u(\alpha) - u(\alpha - 2))$, and $f_y(\beta) = e^{-\beta} u(\beta)$. Find the PDF for z = x + y.*

Solution. We will find f_z using the convolution integral

$$f_z(\gamma) = \int_{-\infty}^{\infty} f_y(\beta) f_x(\gamma - \beta) \, d\beta.$$

It is important to note that the integration variable is β and that γ is constant. For each fixed value of γ the above integral is evaluated by first multiplying $f_y(\beta)$ times $f_x(\gamma - \beta)$ and then finding the area under this product curve. We have

$$f_x(\gamma - \beta) = 0.5(u(\gamma - \beta) - u(\gamma - \beta - 2)).$$

Plots of $f_x(\alpha)$ vs. α and $f_x(\gamma - \beta)$ vs. β, respectively, are shown in Fig. 4.12(a) and (b). The PDF for y is shown in Fig. 4.12(c). Note that Fig. 4.12(b) is obtained from Fig. 4.12(a) by flipping the latter about the $\alpha = 0$ axis and relabeling the origin as γ. Now the integration limits for the desired convolution can easily be obtained by superimposing Fig. 4.12(b) onto Fig. 4.12(c)—the value of γ can be read from the β axis.

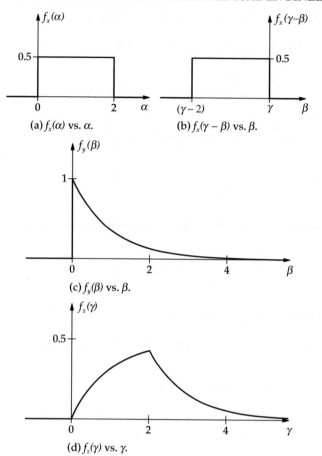

FIGURE 4.12: Plots for Example 4.4.1.

For $\gamma < 0$, we have $f_x(\gamma - \beta) f_y(\beta) = 0$ for all β; hence, $f_z(\gamma) = 0$ for $\gamma < 0$.
For $0 < \gamma < 2$,

$$f_z(\gamma) = \int\limits_{-\infty}^{\gamma} 0.5 e^{-\beta}\, d\beta = 0.5(1 - e^{-\gamma}).$$

For $2 < \gamma$,

$$f_z(\gamma) = \int\limits_{\gamma-2}^{\gamma} 0.5 e^{-\beta}\, d\beta = 0.5 e^{-\gamma}(e^2 - 1).$$

Since the integrand is a bounded function, the resulting f_z is continuous; hence, we know that at the boundaries of the above regions, the results must agree. The result is

$$f_z(\gamma) = \begin{cases} 0, & \gamma \leq 0 \\ 0.5(1 - e^{-\gamma}), & 0 \leq \gamma \leq 2 \\ 0.5e^{-\gamma}(e^2 - 1), & 2 \leq \gamma. \end{cases}$$

The result is shown in Fig. 12.2(d).

$\blacksquare$

One strong motivation for studying Fourier transforms is the fact that the Fourier transform of a convolution is a product of Fourier transforms. The following theorem justifies this statement.

Theorem 4.4.2. *Let F_i be a CDF and*

$$\phi_i(t) = \int_{-\infty}^{\infty} e^{j\alpha t} \, dF_i(\alpha), \qquad (4.66)$$

for $i = 1, 2, 3$. The CDF F_3 may be expressed as the convolution

$$F_3(\gamma) = \int_{-\infty}^{\infty} F_1(\gamma - \beta) \, dF_2(\beta) \qquad (4.67)$$

iff $\phi_3(t) = \phi_1(t)\phi_2(t)$ for all real t.

Proof. Suppose F_3 is given by the above convolution. Let x and y be independent RVs with CDFs F_1 and F_2, respectively. Then $z = x + y$ has CDF F_3 and characteristic function $\phi_3 = \phi_1\phi_2$.

Now suppose that $\phi_3 = \phi_1\phi_2$. Then there exist independent RVs x and y with characteristic functions ϕ_1 and ϕ_2 and corresponding CDFs F_1 and F_2. The RV $z = x + y$ then has characteristic function ϕ_3, and CDF F_3 given by the above convolution. $\blacksquare$

It is important to note that $\phi_{x+y} = \phi_x\phi_y$ is not sufficient to conclude that the RVs x and y are independent. The following example is based on [4, p. 267].

Example 4.4.2. *The RVs x and y have joint PDF*

$$f_{x,y}(\alpha, \beta) = \begin{cases} 0.25(1 + \alpha\beta(\alpha^2 - \beta^2)), & |\alpha| \leq 1, |\beta| \leq 1 \\ 0, & \text{otherwise.} \end{cases}$$

Find: (a) f_x and f_y, (b)ϕ_x and ϕ_y, (c)ϕ_{x+y}, (d)f_{x+y}.

Solution. (a) We have

$$f_x(\alpha) = \frac{1}{4} \int_{-1}^{1} (1 + \alpha\beta(\alpha^2 - \beta^2))\, d\beta = \begin{cases} 0.5, & |\alpha| \leq 1 \\ 0, & \text{otherwise.} \end{cases}$$

Similarly,

$$f_y(\beta) = \frac{1}{4} \int_{-1}^{1} (1 + \alpha\beta(\alpha^2 - \beta^2))\, d\alpha = \begin{cases} 0.5, & |\beta| \leq 1 \\ 0, & \text{otherwise.} \end{cases}$$

(b) From (a) we have

$$\phi_x(t) = \phi_y(t) = \frac{1}{2} \int_{-1}^{1} e^{j\alpha t}\, d\alpha = \frac{\sin t}{t}.$$

(c) We have

$$\phi_{x+y}(t) = \frac{1}{4} \int_{-1}^{1} \int_{-1}^{1} e^{j\alpha t} e^{j\beta t}\, d\alpha\, d\beta + I,$$

where

$$I = \frac{1}{4} \int_{-1}^{1} \int_{-1}^{1} \alpha\beta(\alpha^2 - \beta^2) e^{j\alpha t} e^{j\beta t}\, d\alpha\, d\beta.$$

Interchanging α and β and the order of integration, we obtain

$$I = \frac{1}{4} \int_{-1}^{1} \int_{-1}^{1} \beta\alpha(\beta^2 - \alpha^2) e^{j\beta t} e^{j\alpha t}\, d\alpha\, d\beta = -I.$$

Hence, $I = 0$ and

$$\phi_{x+y}(t) = \left(\frac{\sin t}{t}\right)^2,$$

so that $\phi_{x+y} = \phi_x\phi_y$ even though x and y are not independent. (d) Since $\phi_{x+y} = \phi_x\phi_y$ we have

$$f_{x+y}(\gamma) = \int_{-\infty}^{\infty} f_x(\gamma - \beta) f_y(\beta)\, d\beta.$$

For $-1 < \gamma + 1 < 1$ we find

$$f_{x+y}(\gamma) = \int_{-1}^{\gamma+1} \frac{1}{4} \, d\beta = \frac{\gamma + 2}{4}.$$

For $-1 < \gamma - 1 < 1$ we find

$$f_{x+y}(\gamma) = \int_{\gamma-1}^{1} \frac{1}{4} \, d\beta = \frac{2 - \gamma}{4}.$$

Hence

$$f_{x+y}(\gamma) = \begin{cases} (2 - |\gamma|)/4, & |\gamma| \leq 2 \\ 0, & \text{otherwise.} \end{cases}$$

$\blacksquare$

Drill Problem 4.4.1.   *Random variables x and y have joint PDF*

$$f_{x,y}(\alpha, \beta) = \begin{cases} 4\alpha\beta, & 0 < \alpha < 1, 0 < \beta < 1 \\ 0, & \text{otherwise.} \end{cases}$$

Random variable $z = x + y$. Using convolution, determine: (a) $f_z(-0.5)$, (b) $f_z(0.5)$, (c) $f_z(1.5)$, and (d) $f_z(2.5)$.

Answers: 1/12, 0, 13/12, 0.

4.5   CONDITIONAL PROBABILITY

We previously defined the conditional CDF for the RV x, given event A, as

$$F_{x|A}(\alpha|A) = \frac{P(\zeta \in S : x(\zeta) \leq \alpha, \zeta \in A)}{P(A)}, \qquad (4.68)$$

provided that $P(A) \neq 0$. The extension of this concept to bivariate random variables is immediate:

$$F_{x,y|A}(\alpha, \beta|A) = \frac{P(\zeta \in S : x(\zeta) \leq \alpha, y(\zeta) \leq \beta, \zeta \in A)}{P(A)}, \qquad (4.69)$$

provided that $P(A) \neq 0$.

In this section, we extend this notion to the conditioning event $A = \{\zeta : y(\zeta) = \beta\}$. Clearly, when the RV y is continuous, $P(y = \beta) = 0$, so that some kind of limiting operation is needed.

Definition 4.5.1. *The* **conditional CDF** *for* x, *given* $y = \beta$, *is*

$$F_{x|y}(\alpha \mid \beta) = \lim_{h \to 0} \frac{F_{x,y}(\alpha, \beta) - F_{x,y}(\alpha, \beta - h)}{F_y(\beta) - F_y(\beta - h)}, \qquad (4.70)$$

where the limit is through positive values of h. It is convenient to extend the definition so that $F_{x|y}(\alpha \mid \beta)$ is a legitimate CDF (as a function of α) for any fixed value of β.

Theorem 4.5.1. *Let x and y be jointly distributed RVs.*

If x and y are both discrete RVs then the **conditional PMF** *for x, given $y = \beta$, is*

$$p_{x|y}(\alpha \mid \beta) = \frac{p_{x,y}(\alpha, \beta)}{p_y(\beta)}, \qquad (4.71)$$

for $p_y(\beta) \neq 0$.
If y is a continuous RV then

$$F_{x|y}(\alpha \mid \beta) = \frac{1}{f_y(\beta)} \frac{\partial F_{x,y}(\alpha, \beta)}{\partial \beta}, \qquad (4.72)$$

for $f_y(\beta) \neq 0$.
If x and y are both continuous RVs then the **conditional PDF** *for x, given $y = \beta$ is*

$$f_{x|y}(\alpha \mid \beta) = \frac{f_{x,y}(\alpha, \beta)}{f_y(\beta)}, \qquad (4.73)$$

for $f_y(\beta) \neq 0$.

Proof.   The desired results are a direct consequence of the definitions of CDF, PMF, and PDF.   ∎

Theorem 4.5.2. *Let x and y be independent RVs. Then for all real α,*

$$F_{x|y}(\alpha \mid \beta) = F_x(\alpha). \qquad (4.74)$$

If x and y are discrete independent RVs then for all real α,

$$p_{x|y}(\alpha \mid \beta) = p_x(\alpha). \qquad (4.75)$$

If x and y are continuous independent RVs then for all real α,

$$f_{x|y}(\alpha \mid \beta) = f_x(\alpha). \qquad (4.76)$$

Example 4.5.1. *Random variables x and y have the joint PMF shown in Fig. 4.5. (a) Find the conditional PMF $p_{x,y|A}(\alpha, \beta \mid A)$, if $A = \{\zeta \in S : x(\zeta) \neq y(\zeta)\}$. (b) Find the PMF $p_{y|x}(\beta \mid 1)$. (c) Are x and y conditionally independent, given event $B = \{x < 0\}$?*

Solution. (a) We find

$$P(A^c) = P(x \neq y) = p_{x,y}(1, 1) + p_{x,y}(3, 3) = \frac{1}{4};$$

hence, $P(A) = 1 - P(A^c) = 3/4$. Let $D_{x,y}$ denote the support set for the PMF $p_{x,y}$. Then

$$p_{x,y|A}(\alpha, \beta \mid A) = \begin{cases} \dfrac{p_{x,y}(\alpha, \beta)}{P(A)}, & (\alpha, \beta) \in D_{x,y} \cap \{\alpha \neq \beta\} \\ 0, & \text{otherwise.} \end{cases}$$

The result is shown in graphical form in Fig. 4.13.

(b) We have

$$p_{y|x}(\beta \mid 1) = \frac{p_{x,y}(1, \beta)}{p_x(1)},$$

and

$$p_x(1) = \sum_{\beta} P_{x,y}(1, \beta) = P_{x,y}(1, -1) + P_{x,y}(1, 1) = \frac{3}{8}.$$

Consequently,

$$p_{x|y}(\beta \mid 1) = \begin{cases} 1/3, & \beta = 1 \\ 2/3, & \beta = -1 \\ 0, & \text{otherwise.} \end{cases}$$

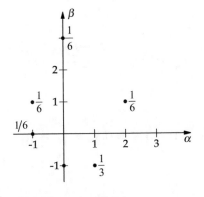

FIGURE 4.13: Conditional PMF for Example 4.5.1a.

(c) The support set for $p_{x,y}$ is $\{(-1, 0), (-1, 1)\}$, and we find easily that $P(B) = 1/4$. Then

$$p_{x,y|B}(\alpha, \beta \,|\, B) = \begin{cases} 0.5, & (\alpha, \beta) = (-1, 0) \\ 0.5, & (\alpha, \beta) = (-1, 1) \\ 0, & \text{otherwise.} \end{cases}$$

Thus

$$p_{x|B}(\alpha \,|\, B) = \begin{cases} 1, & \alpha = -1 \\ 0, & \text{otherwise,} \end{cases}$$

and

$$p_{y|B}(\beta \,|\, B) = \begin{cases} 0.5, & \beta = 0 \\ 0.5, & \beta = 1 \\ 0, & \text{otherwise.} \end{cases}$$

We conclude that x and y are conditionally independent, given B. ■

Example 4.5.2. *Random variables x and y have joint PDF*

$$f_{x,y}(\alpha, \beta) = \begin{cases} 0.25\alpha(1 + 3\beta^2), & 0 < \alpha < 2, 0 < \beta < 1 \\ 0, & \text{otherwise.} \end{cases}$$

Find (a) $P(0 < x < 1 | y = 0.5)$ and (b) $f_{x,y|A}(\alpha, \beta \,|\, A)$, where event $A = \{x + y \leq 1\}$.

Solution. (a) First we find

$$f_y(0.5) = \int_0^2 \frac{\alpha}{4} \frac{7}{4} \, d\alpha = \frac{7}{8}.$$

Then for $0 < \alpha < 2$,

$$f_{x|y}(\alpha \,|\, 0.5) = \frac{0.25\alpha 7/4}{7/8} = \frac{\alpha}{2},$$

and

$$P(0 < x < 1 | y = 0.5) = \int_1^0 \frac{\alpha}{2} d\alpha = \frac{1}{4}.$$

(b) First, we find

$$P(A) = \int_{\alpha+\beta\leq 1} \int f_{x,y}(\alpha, \beta) d\alpha d\beta;$$

substituting,

$$P(A) = \int\limits_0^1 \int\limits_0^{1-\beta} \frac{\alpha}{4}(1 + 3\beta^2)\, d\alpha\, d\beta = \frac{13}{240}.$$

The support region for the PDF $f_{x,y}$ is

$$R = \{(\alpha, \beta) : 0 < \alpha < 2, 0 < \beta < 1\}.$$

Let $B = R \cap \{\alpha + \beta \leq 1\}$. For all $(\alpha, \beta) \in B$, we have

$$f_{x,y|A}(\alpha, \beta \mid A) = \frac{f_{x,y}(\alpha, \beta)}{P(A)} = \frac{60}{13}\alpha(1 + 3\beta^2),$$

and $f_{x,y|A}(\alpha, \beta \mid A) = 0$, otherwise. We note that

$$B = \{(\alpha, \beta) : 0 < \alpha \leq 1 - \beta < 1\}. \qquad \blacksquare$$

Example 4.5.3. *Random variables x and y have joint PDF*

$$f_{x,y}(\alpha, \beta) = \begin{cases} 6\alpha, & 0 < \alpha < 1 - \beta < 1 \\ 0, & \text{otherwise.} \end{cases}$$

Determine whether or not x and y are conditionally independent, given $A = \{\zeta \in S : x \geq y\}$.

Solution. The support region for $f_{x,y}$ is

$$R = \{(\alpha, \beta) : 0 < \alpha < 1 - \beta < 1\};$$

the support region for $f_{x,y|A}$ is thus

$$B = \{(\alpha, \beta) : 0 < \alpha < 1 - \beta < 1, \alpha \geq \beta\} = \{0 < \beta \leq \alpha < 1 - \beta < 1\}.$$

The support regions are illustrated in Fig. 4.14.

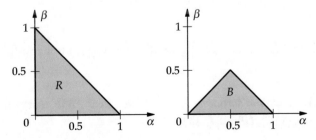

FIGURE 4.14: Support regions for Example 4.5.3.

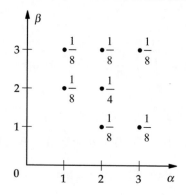

FIGURE 4.15: PMF for Drill Problem 4.5.1.

For $(\alpha, \beta) \in B$,

$$f_{x,y|A}(\alpha, \beta \mid A) = \frac{f_{x,y}(\alpha, \beta)}{P(A)} = \frac{6\alpha}{P(A)}.$$

The conditional marginal densities are found by integrating $f_{x,y|A}$:
For $0 < \beta < 0.5$,

$$f_{y|A}(\beta \mid A) = \frac{1}{P(A)} \int_{\beta}^{1-\beta} 6\alpha \, d\alpha = \frac{3(1 - 2\beta)}{P(A)}.$$

For $0 < \alpha < 0.5$,

$$f_{x|A}(\alpha \mid A) = \frac{1}{P(A)} \int_{0}^{\alpha} 6\alpha \, d\beta = \frac{6\alpha^2}{P(A)}.$$

For $0.5 < \alpha < 1$,

$$f_{x|A}(\alpha \mid A) = \frac{1}{P(A)} \int_{0}^{1-\alpha} 6\alpha \, d\alpha = \frac{6\alpha(1 - \alpha)}{P(A)}.$$

We conclude that since $P(A)$ is a constant, the RVs x and y are not conditionally independent, given A. ∎

Drill Problem 4.5.1. *Random variables x and y have joint PMF shown in Fig. 4.15. Find (a)* $p_x(1)$, *(b)* $p_y(2)$, *(c)* $p_{x|y}(1 \mid 2)$, *(d)* $p_{y|x}(3 \mid 1)$.

Answers: 1/2, 1/4, 3/8, 1/3.

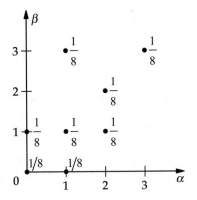

FIGURE 4.16: PMF for Drill Problem 4.5.2.

Drill Problem 4.5.2. *Random variables x and y have joint PMF shown in Fig. 4.16. Event $A = \{\zeta \in S : x + y \leq 1\}$. Find (a) $P(A)$, (b) $p_{x,y|A}(1, 1 \mid A)$, (c) $p_{x|A}(1 \mid A)$, and (d) $p_{y|A}(1 \mid A)$.*

Answers: 0, 3/8, 1/3, 1/3.

Drill Problem 4.5.3. *Random variables x and y have joint PMF shown in Fig. 4.17. Determine if random variables x and y are: (a) independent, (b) independent, given $\{y \leq 1\}$.*

Answers: No, No.

Drill Problem 4.5.4. *The joint PDF for the RVs x and y is*

$$f_{x,y}(\alpha, \beta) = \begin{cases} \dfrac{2}{9}\alpha^2\beta, & 0 < \alpha < 3, 0 < \beta < 1 \\ 0, & \text{otherwise.} \end{cases}$$

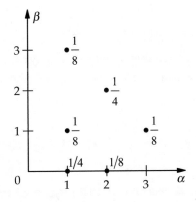

FIGURE 4.17: PMF for Drill Problem 4.5.3.

Find: (a) $f_{x|y}(1|0.5)$, *(b)* $f_{y|x}(0.5|1)$, *(c)* $P(x \leq 1, y \leq 0.5 | x + y \leq 1)$, *and (d)* $P(x \leq 1 | x + y \leq 1)$.

Answers: 1/9, 1, 1, 13/16.

Drill Problem 4.5.5. *The joint PDF for the RVs x and y is*

$$f_{x,y}(\alpha, \beta) = e^{-\alpha - \beta} u(\alpha) u(\beta).$$

Find: (a) $f_{x|y}(1|1)$, *(b)* $F_{x|y}(1|1)$, *(c)* $P(x \geq 5 | x \geq 1)$, *and (d)* $P(x \leq 0.5 | x + y \leq 1)$.

Answers: $1 - e^{-1}$, e^{-1}, e^{-4}, $\dfrac{1 - e^{-0.5} - 0.5e^{-1}}{1 - 2e^{-1}}$.

Drill Problem 4.5.6. *The joint PDF for the RVs x and y is*

$$f_{x,y}(\alpha, \beta) = \begin{cases} \dfrac{2\beta}{\alpha}, & 0 < \beta < \sqrt{\alpha} < 1 \\ 0, & \text{otherwise.} \end{cases}$$

Find: (a) $P(y \leq 0.25 | x = 0.25)$, *(b)* $P(y = 0.25 | x = 0.25)$, *(c)* $P(x \leq 0.25 | x \leq 0.5)$, *and (d)* $P(x \leq 0.25 | x + y \leq 1)$.

Answers: 0, 1/2, 1/4, 0.46695.

Drill Problem 4.5.7. *The joint PDF for the RVs x and y is*

$$f_{x,y}(\alpha, \beta) = \begin{cases} 4\alpha\beta, & 0 < \alpha < 1, 0 < \beta < 1 \\ 0, & \text{otherwise.} \end{cases}$$

Determine whether or not x and y are (a) independent, (b) independent, given $A = \{x + y \geq 1\}$.

Answers: No, Yes.

4.6   CONDITIONAL EXPECTATION

Conditional expectation is completely analogous to ordinary expectation, with the unconditional CDF replaced with the conditional version. In particular, the conditional expectation of $g(x, y)$, given event A, is defined as

$$E(g(x, y) | A) = \int_{-\infty}^{\infty} \int_{-\infty}^{\infty} g(\alpha, \beta) \, dF_{x,y|A}(\alpha, \beta | A). \tag{4.77}$$

When the conditioning event A has zero probability, as when $A = \{x = 0\}$ for continuous RVs, the conditional CDF, PMF, and PDF definitions of the previous sections are used.

Definition 4.6.1. *The **conditional expectation** of $g(x, y)$, given $y = \beta$, is defined by*

$$E(g(x, y) \mid y = \beta) = \int_{-\infty}^{\infty} g(\alpha, \beta) \, dF_{x \mid y}(\alpha \mid \beta). \qquad (4.78)$$

In particular, the conditional mean of x, given $y = \beta$, is

$$E(x \mid y = \beta) = \int_{-\infty}^{\infty} \alpha \, dF_{x \mid y}(\alpha \mid \beta) = \int_{-\infty}^{\infty} \alpha f_{x \mid y}(\alpha \mid \beta) \, d\alpha. \qquad (4.79)$$

It is important to note that if the given value of $y(\beta)$ is a constant, then $E(x \mid y = \beta)$ is also a constant. In general, $E(x \mid y = \beta)$ is a function of β. Once this function is obtained, one may substitute $\beta = y(\zeta)$ and treat the result as a **random variable**; we denote this result as simply $E(x \mid y)$. It is also important to note that conditional expectation, as ordinary expectation, is a linear operation.

Definition 4.6.2. *The **conditional mean** of x, given $y = \beta$, is defined by*

$$\eta_{x \mid y = \beta} = E(x \mid y = \beta), \qquad (4.80)$$

*note that the RV $\eta_{x \mid y} = E(x \mid y)$. The **conditional variance** of x, given $y = \beta$, is defined by*

$$\sigma^2_{x \mid y = \beta} = E((x - \eta_{x \mid y})^2 \mid y = \beta). \qquad (4.81)$$

The RV $\sigma^2_{x \mid y} = E((x - \eta_{x \mid y})^2 \mid y)$.

Example 4.6.1. *Random variable y has PMF*

$$p_y(\alpha) = \begin{cases} 0.25, & \alpha = 1 \\ 0.5, & \alpha = 2 \\ 0.25, & \alpha = 3 \\ 0, & \text{otherwise.} \end{cases}$$

Find the variance of y, given event $A = \{y \text{ odd}\}$.

Solution. We easily find $P(A) = 0.5$ so that

$$p_{y \mid A}(\alpha \mid A) = \begin{cases} 0.5, & \alpha = 1 \\ 0.5, & \alpha = 3 \\ 0, & \text{otherwise.} \end{cases}$$

Then

$$\eta_{y|A} = E(y|A) = \sum_\alpha \alpha^2 p_{y|a}(\alpha|A) = 2$$

and

$$E(y^2|A) = \sum_\alpha \alpha^2 p_{y|A}(\alpha|A) = 5.$$

Finally,

$$\sigma^2_{y|A} = E(y^2|A) - E^2(y|A) = 5 - 4 = 1.$$

Example 4.6.2.   *Random variables x and y have joint PDF*

$$f_{x,y}(\alpha, \beta) = \begin{cases} 2, & \alpha > 0, 0 < \beta < 1 - \alpha \\ 0, & \text{otherwise.} \end{cases}$$

Find $\eta_{x|y} = E(x|y)$, $E(\eta_{x|y})$, and $E(x)$.

Solution.   We first find the marginal PDF

$$f_y(\beta) = \int_{-\infty}^{\infty} f_{x,y}(\alpha, \beta) \, d\alpha = \int_0^{1-\beta} 2 \, d\alpha = 2(1 - \beta),$$

for $0 < \beta < 1$. Then for $0 < \beta < 1$,

$$f_{x|y}(\alpha \,|\, \beta) = \frac{f_{x,y}(\alpha, \beta)}{f_y(\beta)} = \begin{cases} \dfrac{1}{1 - \beta}, & 0 < \alpha < 1 - \beta \\ 0, & \text{otherwise.} \end{cases}$$

Hence, for $0 < \beta < 1$,

$$E(x \,|\, y = \beta) = \int_0^{1-\beta} \frac{\alpha}{1 - \beta} \, d\alpha = \frac{1 - \beta}{2}.$$

We conclude that

$$\eta_{x|y} = E(x \,|\, y) = \frac{1 - y}{2}.$$

Now,

$$E(\eta_{x|y}) = E\left(\frac{1 - y}{2}\right) = \int_0^1 \frac{1 - \beta}{2} 2(1 - \beta) d\beta = \frac{1}{3} = E(x).$$

Example 4.6.3. *Find the conditional variance of y, given $A = \{x \le 0.75\}$, where*

$$f_{x,y}(\alpha, \beta) = 1.5(\alpha^2 + \beta^2)(u(\alpha) - u(\alpha - 1))(u(\beta) - u(\beta - 1)).$$

Solution. First, we find

$$P(A) = \int_0^1 \int_0^{0.75} 1.5(\alpha^2 + \beta^2)\, d\alpha\, d\beta = \frac{75}{128},$$

so that

$$f_{x,y|A}(\alpha, \beta \mid A) = \frac{f_{x,y}(\alpha, \beta)}{P(A)} = \begin{cases} \frac{64}{25}(\alpha^2 + \beta^2), & 0 < \alpha < 0.75, 0 < \beta < 1 \\ 0, & \text{otherwise.} \end{cases}$$

Then for $0 < \beta < 1$,

$$f_{y|A}(\beta \mid A) = \int_0^{0.75} \frac{64}{25}(\alpha^2 + \beta^2)\, d\alpha = \frac{9}{25} + \frac{48}{25}\beta^2.$$

Consequently,

$$E(y \mid A) = \int_0^1 \beta \left(\frac{9}{25} + \frac{48}{25}\beta^2 \right) d\alpha = \frac{66}{100},$$

and

$$E(y^2 \mid A) = \int_0^1 \beta^2 \left(\frac{9}{25} + \frac{48}{25}\beta^2 \right) d\alpha = \frac{378}{750}.$$

Finally,

$$\sigma_{y|A}^2 = E(y^2 \mid A) - E^2(y \mid A) = \frac{513}{7500}. \qquad \blacksquare$$

There are many applications of conditional expectation. One important use is to simplify calculations involving expectation, as by applying the following theorem.

Theorem 4.6.1. *Let x and y be jointly distributed RVs. Then*

$$E(g(x, y)) = E(E(g(x, y) \mid y)) \qquad (4.82)$$

Proof. Note that

$$dF_{x,y}(\alpha, \beta) = dF_{x|y}(\alpha \mid \beta)\, dF_y(\beta).$$

Special cases of this are

$$f_{x,y}(\alpha, \beta) = f_{x|y}(\alpha \,|\, \beta) \, f_y(\beta)$$

and

$$p_{x,y}(\alpha, \beta) = p_{x|y}(\alpha \,|\, \beta) \, p_y(\beta).$$

We thus have

$$E(g(x, y)) = \int\limits_{-\infty}^{\infty} \left(\int\limits_{-\infty}^{\infty} g(\alpha, \beta) \, d F_{x|y}(\alpha \,|\, \beta) \right) dF_y(\beta).$$

Hence

$$E(g(x, y)) = \int\limits_{-\infty}^{\infty} E(g(x, y) \,|\, y = \beta) \, dF_y(\beta),$$

from which the desired result follows. ∎

The conditional mean estimate is one of the most important applications of conditional expectation.

Theorem 4.6.2. *Let x and y be jointly distributed RVs with $\sigma_x^2 < \infty$. The function g which minimizes $E((x - g(y))^2)$ is*

$$g(y) = E(x \,|\, y). \tag{4.83}$$

Proof. We have

$$
\begin{aligned}
E((x - g(y))^2 \,|\, y) &= E((x - \eta_{x|y} + \eta_{x|y} - g(y))^2 \,|\, y) \\
&= \sigma_{x|y}^2 + 2E((x - \eta_{x|y})(\eta_{x|y} - g(y)) \,|\, y) + (\eta_{x|y} - g(y))^2 \\
&= \sigma_{x|y}^2 + (\eta_{x|y} - g(y))^2.
\end{aligned}
$$

The choice $g(y) = \eta_{x|y}$ is thus seen to minimize the above expression, applying the (unconditional) expectation operator yields the desired result. ∎

The above result is extremely important: the best minimum mean square estimate of a quantity is the conditional mean of the quantity, given the data to be used in the estimate. In many cases, the conditional mean is very difficult or even impossible to compute. In the important Gaussian case (discussed in a later chapter) the conditional mean turns out to be easy to find. In fact, in the Gaussian case, the conditional mean is always a linear function of the given data.

Example 4.6.4. *Random variables x and y are independent with*

$$f_x(\alpha) = \begin{cases} 1/20, & |\alpha| \le 10, \\ 0, & \text{otherwise,} \end{cases}$$

and

$$f_y(\beta) = \begin{cases} 1/2, & |\beta| \le 1, \\ 0, & \text{otherwise.} \end{cases}$$

The random variable $z = x + y$. Find (a) $f_z(\gamma)$ and (b) $\hat{x} = g(z)$ to minimize $E((\hat{x} - g(z))^2)$.

Solution. (a) We find f_z using the convolution of f_x with f_y:

$$f_z(\gamma) = \int_{-\infty}^{\infty} f_y(\gamma - \alpha) f_x(\alpha) \, d\alpha.$$

For $-11 < \gamma < -9$,

$$f_z(\gamma) = \int_{-10}^{\gamma+1} \frac{1}{40} \, d\alpha = \frac{\gamma + 11}{40}.$$

For $-9 < \gamma < 9$,

$$f_z(\gamma) = \int_{\gamma-1}^{\gamma+1} \frac{1}{40} \, d\alpha = \frac{1}{20}.$$

For $9 < \gamma < 11$,

$$f_z(\gamma) = \int_{\gamma-1}^{10} \frac{1}{40} \, d\alpha = \frac{11 - \gamma}{40}.$$

Finally, $f_z(\gamma) = 0$ if $|\gamma| > 11$.

(b) From the preceding theorem, we know that $\hat{x} = g(z) = \eta_{x|z}$. Using the fact that $f_{x,z}(\alpha, \gamma) = f_x(\alpha) f_y(\gamma - \alpha)$, we find

$$f_{x|z}(\alpha \mid \gamma) = \frac{f_x(\alpha) f_y(\gamma - \alpha)}{f_z(\gamma)} = \begin{cases} \dfrac{1}{\gamma + 11}, & -10 < \alpha < \gamma + 1, \\[2mm] \dfrac{1}{2}, & \gamma - 1 < \alpha < \gamma + 1, \\[2mm] \dfrac{1}{11 - \gamma}, & \gamma - 1 < \alpha < 10. \end{cases}$$

Notes that for each fixed value of γ with $|\gamma| < 11$, we have that $f_{x|z}(\alpha \,|\, \gamma)$ is a valid PDF (as a function of α). Consequently,

$$E(x|z = \gamma) = \int_{-\infty}^{\infty} \alpha f_{x|z}(\alpha \,|\, \gamma)\, d\alpha$$

$$= \begin{cases} \dfrac{(\gamma + 1)^2 - 100}{2(\gamma + 11)}, & -11 < \gamma < -9, \\[2mm] \dfrac{(\gamma + 1)^2 - (\gamma - 1)^2}{4} = \gamma, & |\gamma| < 9, \\[2mm] \dfrac{100 - (\gamma - 1)^2}{2(11 - \gamma)}, & 9 < \gamma < 11. \end{cases}$$

We conclude that

$$\hat{x} = g(z) = \begin{cases} \dfrac{(z + 1)^2 - 100}{2(z + 11)}, & -11 < z < -9, \\[2mm] z, & |\gamma| < 9, \\[2mm] \dfrac{100 - (z - 1)^2}{2(11 - z)}, & 9 < z < 11. \end{cases}$$

■

Drill Problem 4.6.1. *Random variables x and y have joint PMF shown in Fig. 4.18. Find (a) $E(x \,|\, y = 3)$, (b) $\sigma^2_{x|y=2}$, and (c) $\sigma_{x,y|x+y\geq5}$.*

Answers: 24/25, −3/16, 2.

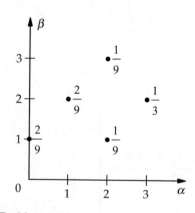

FIGURE 4.18: PMF for Drill Problem 4.6.1.

Drill Problem 4.6.2. *The joint PDF for the RVs x and y is*

$$f_{x,y}(\alpha, \beta) = \begin{cases} \dfrac{2}{9}\alpha^2\beta, & 0 < \alpha < 3, 0 < \beta < 1 \\ 0, & \text{otherwise,} \end{cases}$$

and event $A = \{x + y \le 1\}$. Find: (a) $E(x \,|\, y = 0.5)$, (b) $E(x \,|\, A)$, and (c) $\sigma_{x,y|A}$.

Answers: 9/4, −1/42, 1/2.

Drill Problem 4.6.3. *The joint PDF for the RVs x and y is*

$$f_{x,y}(\alpha, \beta) = \begin{cases} \dfrac{2\beta}{\alpha}, & 0 < \beta < \sqrt{\alpha} < 1 \\ 0, & \text{otherwise.} \end{cases}$$

Determine: (a) $E(y \,|\, x = 0.25)$, (b) $E(x \,|\, x + y \le 1)$, (c) $E(4x - 2 \,|\, x + y \le 1)$, and (d) $\sigma^2_{y|x=0.25}$.

Answers: −0.86732, 1/72, 0.28317, 1/3.

Drill Problem 4.6.4. *The joint PDF for the RVs x and y is*

$$f_{x,y}(\alpha, \beta) = \begin{cases} 4\alpha\beta, & 0 < \alpha < 1, 0 < \beta < 1 \\ 0, & \text{otherwise.} \end{cases}$$

Determine whether or not x and y are (a) independent; (b) independent, given $A = \{x + y \ge 1\}$.

Answers: No, Yes.

4.7   SUMMARY

In this chapter, jointly distributed RVs are considered. The joint CDF for the RVs x and y is defined as

$$F_{x,y}(\alpha, \beta) = P(\zeta \in S : x(\zeta) \le \alpha, y(\zeta) \le \beta). \tag{4.84}$$

Probabilities for rectangular-shaped regions, as well as marginal CDFs are easily obtained directly from the joint CDF. If the RVs x and y are jointly discrete, the joint PMF

$$p_{x,y}(\alpha, \beta) = P(\zeta \in S : x(\zeta) = \alpha, y(\zeta) = \beta) \tag{4.85}$$

can be obtained from the joint CDF, and probabilities can be computed using a two-dimensional summation. If the RVs are jointly continuous (or if Dirac delta functions are permitted) then the joint PDF is defined by

$$f_{x,y}(\alpha, \beta) = \frac{\partial^2 F_{x,y}(\alpha, \beta)}{\partial \beta\, \partial \alpha}, \tag{4.86}$$

where left-hand derivatives are assumed. The two-dimensional Riemann-Stieltjes integral can be applied in the general mixed RV case.

The expectation operator is defined as

$$E(g(x, y)) = \int\limits_{-\infty}^{\infty} g(\alpha, \beta) \, dF_{x,y}(\alpha, \beta). \qquad (4.87)$$

Various moments, along with the moment generating function are defined. The correlation coefficient is related to the covariance and standard deviations by $\rho_{x,y} = \sigma_{x,y}/(\sigma_x \sigma_y)$, and is seen to satisfy $|\rho_{x,y}| \leq 1$. Some important inequalities are presented. The two-dimensional characteristic function is seen to be a straightforward extension of the one–dimensional case.

A convolution operation arises naturally when determining the distribution for the sum of two independent RVs. Characteristic functions provide an alternative method for computing a convolution.

The conditional CDF, given the value of a RV, is defined as

$$F_{x|y}(\alpha \mid \beta) = \lim_{h \to 0} \frac{F_{x,y}(\alpha, \beta) - F_{x,y}(\alpha, \beta - h)}{F_y(\beta) - F_y(\beta - h)}; \qquad (4.88)$$

the corresponding conditional PMF and PDF follow in a straightforward manner. The conditional expectation of x, given $y = \beta$, is defined as

$$E(x \mid y = \beta) = \int\limits_{-\infty}^{\infty} \alpha \, dF_{x|y}(\alpha \mid \beta). \qquad (4.89)$$

As we will see, all of these concepts extend in a logical manner to the n-dimensional case—the extension is aided greatly by the use of vector–matrix notation.

4.8   PROBLEMS

1. Which of the following functions are legitimate PDFs? Why, or why not?
   (a)

$$g_1(\alpha, \beta) = \begin{cases} \alpha^2 + 0.5\alpha\beta, & 0 \leq \alpha \leq 1, 0 \leq \beta \leq 2 \\ 0, & \text{otherwise.} \end{cases}$$

   (b)

$$g_2(\alpha, \beta) = \begin{cases} 2(\alpha + \beta - 2\alpha\beta), & 0 \leq \alpha \leq 1, 0 \leq \beta \leq 1 \\ 0, & \text{otherwise.} \end{cases}$$

(c)

$$g_3(\alpha, \beta) = \begin{cases} e^{-\alpha}e^{-\beta}, & \alpha > 0, \beta > 0 \\ 0, & \text{otherwise.} \end{cases}$$

(d)

$$g_4(\alpha, \beta) = \begin{cases} \alpha \cos(\beta), & 0 \le \alpha \le 1, 0 \le \beta \le \pi \\ 0, & \text{otherwise.} \end{cases}$$

2. Find the CDF $F_{x,y}(\alpha, \beta)$ if

$$f_{x,y}(\alpha, \beta) = \begin{cases} 0.25, & 0 \le \beta \le 2, \beta \le \alpha \le \beta + 2 \\ 0, & \text{otherwise.} \end{cases}$$

3. Random variables x and y have joint PDF

$$f_{x,y}(\alpha, \beta) = \begin{cases} a\alpha^2, & 0 \le \beta \le 1, 1 \le \alpha \le e^\beta \\ 0, & \text{otherwise.} \end{cases}$$

Determine: (a) a, (b) $f_x(\alpha)$, (c) $f_y(\beta)$, (d) $P(x \le 2)$.

4. With the joint PDF of random variables x and y given by

$$f_{x,y}(\alpha, \beta) = \begin{cases} a(\alpha^2 + \beta^2), & -1 < \alpha < 1, 0 < \beta < 2 \\ 0, & \text{otherwise.} \end{cases}$$

Determine: (a) a, (b) $P(-0.5 < x < 0.5, 0 < y < 1)$, (c) $P(-0.5 < x < 0.5)$, (d) $P(|xy| > 1)$.

5. The joint PDF for random variables x and y is

$$f_{x,y}(\alpha, \beta) = \begin{cases} a(\alpha^2 + \beta^2), & 0 < \alpha < 2, 1 < \beta < 4 \\ 0, & \text{otherwise.} \end{cases}$$

Determine: (a) a, (b) $P(1 \le x \le 2, 2 \le y \le 3)$, (c) $P(1 < x < 2)$, (d) $P(x + y > 4)$.

6. Given

$$f_{x,y}(\alpha, \beta) = \begin{cases} a(\alpha^2 + \beta), & 0 < \alpha < 1, 0 < \beta < 1 \\ 0, & \text{otherwise.} \end{cases}$$

Determine: (a) a, (b) $P(0 < x < 1/2, 1/4 < y < 1/2)$, (c) $f_y(\beta)$, (d) $f_x(\alpha)$.

7. The joint PDF for random variables x and y is

$$f_{x,y}(\alpha, \beta) = \begin{cases} a|\alpha\beta|, & |\alpha| < 1, |\beta| < 1 \\ 0, & \text{otherwise.} \end{cases}$$

Determine (a) a, (b) $P(x > 0)$, (c) $P(xy > 0)$, (d) $P(x - y < 0)$.

8. Given

$$f_{x,y}(\alpha, \beta) = \begin{cases} a\dfrac{\beta}{\alpha}, & 0 < \beta < \alpha < 1 \\ 0, & \text{otherwise.} \end{cases}$$

Determine: (a) a, (b) $P(1/2 < x < 1, 0 < y < 1/2)$, (c) $P(x + y < 1)$, (d) $f_x(\alpha)$.

9. The joint PDF for random variables x and y is

$$f_{x,y}(\alpha, \beta) = \begin{cases} \dfrac{1}{50}(\alpha^2 + \beta^2), & 0 < \alpha < 2, 1 < \beta < 4 \\ 0, & \text{otherwise.} \end{cases}$$

Determine: (a) $P(y < 4 | x = 1)$, (b) $P(y < 2 | x = 1)$, (c) $P(y < 3 | x + y > 4)$.

10. Random variables x and y have the following joint PDF.

$$f_{x,y}(\alpha, \beta) = \begin{cases} a\alpha \exp(-\alpha(1 + \beta)), & \alpha > 0, \beta > 0 \\ 0, & \text{otherwise.} \end{cases}$$

Find: (a) a, (b) $f_x(\alpha)$, (c) $f_y(\beta)$, (d) $f_{x|y}(\alpha | \beta)$, (e) $f_{y|x}(\beta | \alpha)$.

11. Random variables x and y have joint PDF

$$f_{x,y}(\alpha, \beta) = \begin{cases} \dfrac{1}{2\alpha^2\beta}, & \alpha \geq 1, \dfrac{1}{\alpha} \leq \beta \leq \alpha \\ 0, & \text{otherwise.} \end{cases}$$

Event $A = \{\max(x, y) \leq 2\}$. Find: (a) $f_{x,y|A}(\alpha, \beta | A)$, (b) $f_{x|A}(\alpha | A)$, (c) $f_{y|A}(\beta | A)$, (d) $f_{x|y}(\alpha | \beta)$, (e) $f_{y|x}(\beta | \alpha)$.

12. Random variables x and y have joint PDF

$$f_{x,y}(\alpha, \beta) = \begin{cases} \dfrac{3}{32}(\alpha^3 + 4\beta), & 0 \leq \alpha \leq 2, \alpha^2 \leq \beta \leq 2\alpha \\ 0, & \text{otherwise.} \end{cases}$$

Event $A = \{y \leq 2\}$. Find: (a) $f_{x,y|A}(\alpha, \beta)$, (b) $f_{x|A}(\alpha | A)$, (c) $f_{y|A}(\beta | A)$, (d) $f_{x|y}(\alpha | \beta)$, (e) $f_{y|x}(\beta | \alpha)$.

13. The joint PDF for random variables x and y is

$$f_{x,y}(\alpha, \beta) = \begin{cases} a, & \alpha^2 < \beta < \alpha \\ 0, & \text{otherwise.} \end{cases}$$

Determine: (a) a, (b) $P(x \leq 1/2, y \leq 1/2)$, (c) $P(x \leq 1/4)$, (d) $P(y < 1/2 - x)$, (e) $P(x < 3/5 \,|\, y = 3/4)$.

14. Random variables x and y have joint PDF

$$f_{x,y}(\alpha, \beta) = \begin{cases} a, & \alpha + \beta \leq 1, 0 \leq \alpha, 0 \leq \beta \\ 0, & \text{otherwise.} \end{cases}$$

Determine: (a) a, (b) $F_{x,y}(\alpha, \beta)$, (c) $P(x < 3/4)$, (d) $P(y < 1/4 \,|\, x \leq 3/4)$, (e) $P(x > y)$.

15. The joint PDF for random variables x and y is

$$f_{x,y}(\alpha, \beta) = \begin{cases} \dfrac{3}{8}\alpha, & 0 \leq \beta \leq \alpha \leq 2 \\ 0, & \text{otherwise.} \end{cases}$$

Event $A = \{x \leq 2 - y\}$. Determine: (a) $f_x(\alpha)$, (b) $f_y(\beta)$, (c) $f_{x|y}(\alpha \,|\, \beta)$, (d) $f_{y|x}(\beta \,|\, \alpha)$, (e) $f_{x|A}(\alpha \,|\, A)$, (f) $f_{y|A}(\beta \,|\, A)$.

16. Random variables x and y have joint PDF

$$f_{x,y}(\alpha, \beta) = \begin{cases} 8\alpha\beta, & 0 \leq \alpha^2 + \beta^2 \leq 1, \alpha \geq 0, \beta \geq 0 \\ 0, & \text{otherwise.} \end{cases}$$

Let event $A = \{x \geq y\}$. Determine: (a) $P(A)$, (b) $f_{x,y|A}(\alpha, \beta \,|\, A)$, (c) $f_{x|A}(\alpha \,|\, A)$.

17. Random variables x and y have joint PDF

$$f_{x,y}(\alpha, \beta) = \begin{cases} \dfrac{1}{8}(\alpha^2 - \beta^2) \exp(-\alpha), & \alpha \geq 0, |\beta| \leq \alpha \\ 0, & \text{otherwise.} \end{cases}$$

(a) Determine $f_{y|x}(\beta \,|\, \alpha)$. (b) Write the integral(s) necessary to find the marginal PDF for y (do not solve). (c) Given the event $B = \{x^2 + y^2 \leq 1\}$, write the integral(s) necessary to find $P(B)$ (do not solve).

18. Random variables x and y have joint PDF

$$f_{x,y}(\alpha, \beta) = \begin{cases} a\alpha^2\beta(2 - \beta), & 0 \leq \alpha \leq 2, 0 \leq \beta \leq 2 \\ 0, & \text{otherwise.} \end{cases}$$

Determine: (a) a, (b) $f_y(\beta)$, (c) $f_{x|y}(\alpha \,|\, \beta)$, (d) whether or not x and y are independent.

19. Given

$$f_{x,y}(\alpha, \beta) = \begin{cases} \dfrac{2}{9}\alpha^2\beta, & 0 < \alpha < 3, 0 < \beta < 1 \\ 0, & \text{otherwise,} \end{cases}$$

and event $A = \{x < y\}$. Determine: (a) $f_{x|y}(\alpha \mid \beta)$; (b) $f_{y|x}(\beta \mid \alpha)$; (c) $P(x < 2 \mid y = 3/4)$; (d) $P(x \leq 1, y \leq 0.5 \mid A)$; (e) $P(y \leq 0.5 \mid A)$; (f) whether or not x and y are independent; (g) whether or not x and y are independent, given A.

20. Determine if random variables x and y are independent if

$$f_{x,y}(\alpha, \beta) = \begin{cases} 0.6(\alpha + \beta^2), & 0 < \alpha < 1, |\beta| < 1 \\ 0, & \text{otherwise.} \end{cases}$$

21. Given

$$f_{x,y}(\alpha, \beta) = \begin{cases} 10\alpha^2\beta, & 0 \leq \beta \leq \alpha \leq 1 \\ 0, & \text{otherwise,} \end{cases}$$

and event $A = \{x + y > 1\}$. Determine: (a) $f_{y|x}(\beta \mid 3/4)$; (b) $f_{y|A}(\beta \mid A)$; (c) whether x and y are independent random variables, given A.

22. The joint PDF for x and y is given by

$$f_{x,y}(\alpha, \beta) = \begin{cases} 2, & 0 < \alpha < \beta < 1 \\ 0, & \text{otherwise.} \end{cases}$$

Event $A = \{1/2 < y < 3/4, 1/2 < x\}$. Determine whether random variables x and y are: (a) independent; (b) conditionally independent, given A.

23. Random variables x and y have joint PDF

$$f_{x,y}(\alpha, \beta) = \begin{cases} 2, & \alpha + \beta \leq 1, \alpha \geq 0, \beta \geq 0 \\ 0, & \text{otherwise.} \end{cases}$$

Are random variables x and y: (a) independent; (b) conditionally independent, given $\max(x, y) \leq 1/2$?

24. Given

$$f_{x,y}(\alpha, \beta) = \begin{cases} 6(1 - \alpha - \beta), & \alpha + \beta \leq 1, \alpha \geq 0, \beta \geq 0 \\ 0, & \text{otherwise.} \end{cases}$$

Determine: (a) $f_{x|y}(\alpha \mid \beta)$, (b) $F_{x|y}(\alpha \mid \beta)$, (c) $P(x < 1/2 \mid y = 1/2)$, (d) $f_{y|x}(\beta \mid \alpha)$, (e) whether x and y are independent.

25. Random variables x and y have joint PDF

$$f_{x,y}(\alpha, \beta) = \begin{cases} \beta \sin(\alpha), & 0 \le \beta \le 1, 0 \le \alpha \le \pi \\ 0, & \text{otherwise.} \end{cases}$$

Event $A = \{y \ge 0.5\}$ and $B = \{x > y\}$. Determine whether random variables x and y are: (a) independent; (b) conditionally independent, given A; (c) conditionally independent, given B.

26. With the joint PDF of random variables x and y given by

$$f_{x,y}(\alpha, \beta) = \begin{cases} a(\alpha^2 + \beta^2), & |\alpha| < 1, 0 < \beta < 2 \\ 0, & \text{otherwise,} \end{cases}$$

determine (a) $f_x(\alpha)$, (b) $f_y(\beta)$, (c) $f_{x|y}(\alpha \,|\, \beta)$, (d) whether x and y are independent.

27. The joint PDF for random variables x and y is

$$f_{x,y}(\alpha, \beta) = \begin{cases} a|\alpha\beta|, & |\alpha| < 1, |\beta| < 1 \\ 0, & \text{otherwise.} \end{cases}$$

Event $A = \{xy > 0\}$. Determine (a) a; (b) $f_{x|A}(\alpha \,|\, A)$; (c) $f_{y|A}(\beta \,|\, A)$; (d) whether x and y are conditionally independent, given A.

28. Let the PDF of random variables x and y be

$$f_{x,y}(\alpha, \beta) = \begin{cases} a\alpha \exp(-(\alpha + \beta)), & \alpha > 0, \beta > 0 \\ 0, & \text{otherwise.} \end{cases}$$

Determine (a) a, (b) $f_x(\alpha)$, (c) $f_y(\beta)$, (d) $f_{x|y}(\alpha \,|\, \beta)$, (e) whether x and y are independent.

29. Given

$$f_{x,y}(\alpha, \beta) = \begin{cases} 6\alpha^2\beta, & 0 < \alpha < 1, 0 < \beta < 1 \\ 0, & \text{otherwise,} \end{cases}$$

and event $A = \{y < x\}$. Determine: (a) $P(0 < x < 1/2, 0 < y < 1/2 \,|\, A)$; (b) $f_{x|A}(\alpha \,|\, A)$; (c) $f_{y|A}(\beta \,|\, A)$; (d) whether x and y are independent, given A.

30. Determine the probability that an experimental value of x will be greater than $E(x)$ if

$$f_{x,y}(\alpha, \beta) = \begin{cases} a(\alpha^2\beta + 1), & \alpha \ge 0, 0 \le \beta \le 2 - 0.5\alpha \\ 0, & \text{otherwise.} \end{cases}$$

31. Random variables x and y have joint PDF

$$f_{x,y}(\alpha, \beta) = \begin{cases} 2, & \alpha + \beta \leq 1, \alpha \geq 0, \beta \geq 0 \\ 0, & \text{otherwise.} \end{cases}$$

Determine: (a) $E(x)$, (b) $E(y \mid x \leq 3/4)$, (c) σ_x^2, (d) $\sigma_{y\mid A}^2$, where $A = \{x \geq y\}$, (e) $\sigma_{x,y}$.

32. The joint PDF for random variables x and y is

$$f_{x,y}(\alpha, \beta) = \begin{cases} 12\alpha(1 - \beta), & \alpha \geq 0, \alpha^2 \leq \beta \leq 1 \\ 0, & \text{otherwise.} \end{cases}$$

Event $A = \{y \geq x^{1/2}\}$. Determine: (a) $E(x)$; (b) $E(y)$; (c) $E(x \mid A)$; (d) $E(y \mid A)$; (e) $E(x + y \mid A)$; (f) $E(x^2 \mid A)$; (g) $E(3x^2 + 4x + 3y \mid A)$; (h) the conditional covariance for x and y, given A; (i) whether x and y are conditionally independent, given A; (j) the conditional variance for x, given A.

33. Suppose x and y have joint PDF

$$f_{x,y}(\alpha, \beta) = \begin{cases} \dfrac{16\beta}{\alpha^3}, & \alpha > 2, 0 < \beta < 1 \\ 0, & \text{otherwise.} \end{cases}$$

Determine: (a) $E(x)$, (b) $E(y)$, (c) $E(xy)$, (d) $\sigma_{x,y}$.

34. The joint PDF of random variables x and y is

$$f_{x,y}(\alpha, \beta) = \begin{cases} a(\alpha + \beta^2), & 0 < \alpha < 1, |\beta| < 1 \\ 0, & \text{otherwise.} \end{cases}$$

Event $A = \{y > x\}$. Determine: (a) a; (b) $f_x(\alpha)$; (c) $f_{y\mid x}(\beta \mid \alpha)$; (d) $E(y \mid x = \alpha)$; (e) $E(xy)$; (f) $f_{x,y\mid A}(\alpha, \beta \mid A)$; (g) $E(x \mid A)$; (h) whether x and y are independent; (i) whether x and y are conditionally independent, given A.

35. Suppose

$$f_x(\alpha) = \frac{\alpha}{8}(u(\alpha) - u(\alpha - 4))$$

and

$$f_{y\mid x}(\beta \mid \alpha) = \begin{cases} 1/\alpha, & 0 \leq \beta \leq \alpha \leq 4 \\ 0, & \text{otherwise.} \end{cases}$$

Determine: (a) $f_{x,y}(\alpha, \beta)$, (b) $f_y(\beta)$, (c) $E(x - y)$, (d) $P(x < 2 \mid y < 2)$, (e) $P(x - y < 1 \mid y < 2)$.

36. The joint PDF of random variables x and y is

$$f_{x,y}(\alpha, \beta) = \begin{cases} a\alpha, & \alpha > 0, -1 < \beta - \alpha < \beta < 0 \\ 0, & \text{otherwise.} \end{cases}$$

Event $A = \{0 > y > -0.5\}$. Determine (a) a, (b) $f_x(\alpha)$, (c) $f_y(\beta)$, (d) $E(x)$, (e) $E(y)$, (f) $E(x^2)$, (g) $E(y^2)$, (h) $E(xy)$, (i) σ_x^2, (j) σ_y^2, (k) $\sigma_{x,y}$, (l) $f_{x,y|A}(\alpha, \beta \,|\, A)$, (m) $E(x \,|\, A)$.

37. Random variables x and y have joint PDF

$$f_{x,y}(\alpha, \beta) = \begin{cases} 0.6(\alpha + \beta^2), & 0 < \alpha < 1, |\beta| < 1 \\ 0, & \text{otherwise.} \end{cases}$$

Determine: (a) $E(x)$, (b) $E(y)$, (c) σ_x^2, (d) σ_y^2, (e) $\sigma_{x,y}$, (f) $E(y \,|\, x = \alpha)$, (g) $E(x \,|\, y = \beta)$, (h) $\sigma_{y|x}^2$, (i) $\sigma_{x|y}^2$.

38. Given

$$f_{x,y}(\alpha, \beta) = \begin{cases} 1.2(\alpha^2 + \beta), & 0 \le \alpha \le 1, 0 \le \beta \le 1 \\ 0, & \text{otherwise.} \end{cases}$$

Event $A = \{y < x\}$. Determine: (a) η_y, (b) $\eta_{x|y=1/2}$, (c) $E(x \,|\, A)$, (d) $\sigma_{x,y}$, (e) $\sigma_{x,y|A}$, (f) $\sigma_{x|y=1/2}^2$, (g) $\sigma_{x|A}^2$.

39. Random variables x and y have joint PDF

$$f_{x,y}(\alpha, \beta) = \begin{cases} \beta \sin(\alpha), & 0 \le \alpha \le \pi, 0 \le \beta \le 1 \\ 0, & \text{otherwise.} \end{cases}$$

Event $A = \{y \ge 0.5\}$ and $B = \{x > y\}$. Determine: (a) $E(x \,|\, A)$, (b) $E(y \,|\, A)$, (c) $E(x \,|\, B)$, (d) $E(y \,|\, B)$, (e) $\rho_{x,y}$, (f) $\rho_{x,y|A}$.

40. If random variables x and y have joint PDF

$$f_{x,y}(\alpha, \beta) = \begin{cases} 0.5\beta \exp(-\alpha), & \alpha \ge 0, 0 \le \beta \le 2 \\ 0, & \text{otherwise,} \end{cases}$$

determine: (a) $\sigma_{x,y}$, (b) $\rho_{x,y}$, (c) $E(y \,|\, x = \alpha)$, (d) $\sigma_{x|y}$.

41. The joint PDF for random variables x and y is

$$f_{x,y}(\alpha, \beta) = \begin{cases} 10\alpha^2\beta, & 0 \le \beta \le \alpha \le 1 \\ 0, & \text{otherwise.} \end{cases}$$

Event $A = \{x + y > 1\}$. Determine: (a) $E(y \,|\, x = 3/4)$, (b) $E(y \,|\, A)$, (c) $E(y^2 \,|\, A)$, (d) $E(5y^2 - 3y + 2 \,|\, A)$, (e) $\sigma_{y|A}^2$, (f) $\sigma_{y|x=3/4}^2$.

42. Random variables x and y have joint PDF

$$f_{x,y}(\alpha, \beta) = \begin{cases} a(\alpha\beta + 1), & 0 < \alpha < 1, 0 < \beta < 1 \\ 0, & \text{otherwise.} \end{cases}$$

Event $A = \{x > y\}$. Find: (a) a, (b) $f_y(\beta)$, (c) $f_{x|y}(\alpha|\beta)$, (d) $E(y)$, (e) $E(x|y)$, (f) $E(xy)$, (g) $P(A)$, (h) $f_{x,y|A}(\alpha, \beta | A)$, (i) $E(xy|A)$.

43. Let random variables x and y have joint PDF

$$f_{x,y}(\alpha, \beta) = \begin{cases} 1/16, & 0 \le \alpha \le 8, |\beta| \le 1 \\ 0, & \text{otherwise.} \end{cases}$$

Random variable $z = yu(y)$. Determine: (a) σ_x, (b) σ_y, (c) σ_z.

44. Random variables x and y have joint PDF

$$f_{x,y}(\alpha, \beta) = \begin{cases} 3(\alpha^2 + \beta^2), & 0 \le \beta \le \alpha \le 1 \\ 0, & \text{otherwise.} \end{cases}$$

Event $A = \{x^2 + y^2 \le 1\}$. Determine: (a) $\sigma_{x,y}$, (b) $\rho_{x,y}$, (c) $\sigma_{x,y|A}$, (d) $\rho_{x,y|A}$.

45. The joint PDF for random variables x and y is

$$f_{x,y}(\alpha, \beta) = \begin{cases} \dfrac{9}{208}\alpha^2\beta^2, & 0 \le \beta \le 2, 1 \le \alpha \le 3 \\ 0, & \text{otherwise.} \end{cases}$$

Determine: (a) σ_x^2, (b) $E(x|y)$, (c) whether x and y are independent, (d) $E(g(x))$ if $g(x) = 26\sin(\pi x)/3$, (e) $E(h(x, y))$ if $h(x, y) = xy$.

46. Suppose random variables x and y are independent with

$$f_{x,y}(\alpha, \beta) = \begin{cases} 2\exp(-2\alpha), & \alpha > 0, 0 \le \beta \le 1 \\ 0, & \text{otherwise.} \end{cases}$$

Determine $E(y(x + y))$.

47. Prove the following properties: (a) Given random variable x and constants a and b, $E(ax + b) = aE(x) + b$. (b) Given independent random variables x and y, $E(xy) = E(x)E(y)$. (c) Given random variable x, constants a and b, and an event A, $E(ax + b | A) = aE(x|A) + b$. (d) Given that random variables x and y are conditionally independent, given event A, $E(xy|A) = E(x|A)E(y|A)$.

48. Random variables x and y have the joint PDF

$$f_{x,y}(\alpha, \beta) = \frac{1}{4}(u(\alpha) - u(\alpha - 2))(u(\beta) - u(\beta - 2)).$$

If $z = x + y$, use convolution to find f_z.

49. Random variables x and y are independent with

$$f_x(\alpha) = e^{-\alpha} u(\alpha)$$

and

$$f_y(\beta) = 2e^{-2\beta} u(\beta).$$

If $z = x + y$, use convolution to find f_z.

50. Independent random variables x and y have PDFs

$$f_x(\alpha) = 2e^{-2\alpha} u(\alpha)$$

and

$$f_y(\beta) = \frac{1}{2}(u(\beta + 1) - u(\beta - 1)).$$

Find f_z if $z = x + y$. Use convolution.

51. Random variables x and y are independent and RV $z = x + y$. Given

$$f_x(\alpha) = u(\alpha - 1) - u(\alpha - 2)$$

and

$$f_y(\beta) = \frac{1}{\sqrt{2}}(u(\beta) - u(\beta - \sqrt{2})),$$

use convolution to find f_z.

52. Random variables x and y are independent with

$$f_x(\alpha) = 2e^{-2\alpha} u(\alpha)$$

and

$$f_y(\beta) = \frac{1}{2}(u(\beta + 1) - u(\beta - 1)).$$

With $z = x + y$, use the characteristic function to find f_z.

53. An urn contains four balls labeled 1, 2, 3, and 4. An experiment involves draw-
ing three balls one after the other without replacement. Let RV x denote the sum
of numbers on first two balls minus the number on the third. Let RV y denote
the product of the numbers on the first two balls minus the number on the third.
Event $A = \{$either x or y is negative$\}$. Determine: (a) $p_{x,y}(\alpha, \beta)$; (b) $p_x(\alpha)$; (c) $p_y(\beta)$;
(d) $p_{y|x}(\beta \mid 5)$; (e) $p_{x|y}(\alpha \mid 5)$; (f) $E(y \mid x = 5)$; (g) $\sigma^2_{y|x=5}$; (h) $p_{x,y|A}(\alpha, \beta \mid A)$; (i) $E(x|A)$;
(j) whether or not x and y are independent; (k) whether or not x and y are independent,
given A; (l) $\sigma_{x,y}$; and (m) $\sigma_{x,y|A}$.

TABLE 4.1: Joint PMF for Problems 54–59.

α	$\beta = 0$	$\beta = 1$	$\beta = 2$
0	6/56	18/56	6/56
1	12/56	11/56	1/56
2	1/56	0	1/56

54. Random variables x and y have the joint PMF given in Table 4.1. Event $A = \{x + y \le 2\}$. Determine: (a) p_x; (b) p_y; (c) $p_{x|y}(\alpha \,|\, 0)$; (d) $p_{y|x}(\beta \,|\, 1)$; (e) $p_{x,y|A}(\alpha, \beta \,|\, A)$; (f) $p_{x|A}$; (g) $p_{y|A}$; (h) whether or not x and y are independent; (i) whether or not x and y are independent, given A.

55. Random variables x and y have the joint PMF given in Table 4.1. Event $A = \{x + y \le 2\}$. Determine: (a) $E(x)$, (b) $E(x^2)$, (c) σ_x^2, (d) $E(5x)$, (e) σ_{2x+1}^2, (f) $E(x - 3x^2)$, (g) $E(x \,|\, A)$, (h) $E(x^2 \,|\, A)$, (i) $E(3x^2 - 2x \,|\, A)$.

56. Random variables x and y have the joint PMF given in Table 4.1. Event $A = \{x + y \le 2\}$. Determine: (a) $E(y)$, (b) $E(y^2)$, (c) σ_y^2, (d) $E(5y - 2)$, (e) σ_{3y}^2, (f) $E(5y - y^2)$, (g) $E(y \,|\, A)$, (h) $E(y^2 \,|\, A)$, (i) $E(3y^2 - 2y \,|\, A)$.

57. Random variables x and y have the joint PMF given in Table 4.1. Event $A = \{x + y \le 2\}$. If $w(x, y) = x + y$, then determine: (a) p_w, (b) $p_{w|A}$, (c) $E(w)$, (d) $E(w \,|\, A)$, (e) σ_w^2, (f) $\sigma_{w|A}^2$.

58. Random variables x and y have the joint PMF given in Table 4.1. Event $A = \{x + y \le 2\}$. If $z(x, y) = x^2 - y$, then determine: (a) p_z, (b) $p_{z|A}$, (c) $E(z)$, (d) $E(z \,|\, A)$, (e) σ_z^2, (f) $\sigma_{z|A}^2$.

59. Random variables x and y have the joint PMF given in Table 4.1. Event $B = \{zw > 0\}$, where $w(x, y) = x + y$, and $z(x, y) = x^2 - y$. Determine: (a) $p_{z,w}$, (b) p_z, (c) p_w, (d) $p_{z|w}(\gamma \,|\, 2)$, (e) $p_{z|B}$, (f) η_z, (g) $\eta_{z|B}$, (h) σ_z^2, (i) $\sigma_{z|B}^2$, (j) $\sigma_{z,w}$, (k) $\sigma_{z,w|B}$.

60. Random variables x and y have joint PMF shown in Fig. 4.19. Event $A = \{xy \ge 1\}$. Determine: (a) p_x; (b) p_y; (c) $p_{x|y}(\alpha \,|\, 1)$; (d) $p_{y|x}(\beta \,|\, 1)$; (e) $p_{x,y|A}(\alpha, \beta \,|\, A)$; (f) $p_{x|A}$; (g) $p_{y|A}$; (h) whether or not x and y are independent; (i) whether or not x and y are independent, given A.

61. Random variables x and y have joint PMF shown in Fig. 4.19. Event $A = \{xy \ge 1\}$. Determine: (a) $E(x)$, (b) $E(x^2)$, (c) σ_x^2, (d) $E(x - 1)$, (e) σ_{3x}^2, (f) $E(5x - 3x^2)$, (g) $E(x \,|\, A)$, (h) $E(x^2 \,|\, A)$, (i) $E(x^2 + 2x \,|\, A)$.

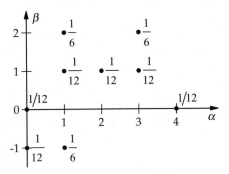

FIGURE 4.19: PMF for Problems 60–65.

62. Random variables x and y have joint PMF shown in Fig. 4.19. Event $A = \{xy \geq 1\}$. Determine: (a) $E(x + y)$, (b) $E(y^2)$, (c) σ_y^2, (d) $E(5y - x)$, (e) $\sigma_{x,y}^2$, (f) $\sigma_{x,y|A}$, (g) $E(x + y \,|\, A)$, (h) $E(x^2 + y^2 \,|\, A)$, (i) $E(3y^2 - 2x \,|\, A)$.

63. Random variables x and y have joint PMF shown in Fig. 4.19. Event $A = \{xy \geq 1\}$. If $w(x, y) = |x - y|$, then determine: (a) p_w, (b) $p_{w|A}$, (c) $E(w)$, (d) $E(w \,|\, A)$, (e) σ_w^2, (f) $\sigma_{w|A}^2$.

64. Random variables x and y have joint PMF shown in Fig. 4.19. Event $A = \{xy \geq 1\}$. If $z(x, y) = 2x - y$, then determine: (a) p_z, (b) $p_{z|A}$, (c) $E(z)$, (d) $E(z \,|\, A)$, (e) σ_z^2, (f) $\sigma_{z|A}^2$.

65. Random variables x and y have joint PMF shown in Fig. 4.19. Event $B = \{z + w \leq 2\}$, where $w(x, y) = |x - y|$, and $z(x, y) = 2x - y$. Determine: (a) $p_{z,w}$, (b) p_z, (c) p_w, (d) $p_{z|w}(\gamma \,|\, 0)$, (e) $p_{z|B}$, (f) η_z, (g) $\eta_{z|B}$, (h) σ_z^2, (i) $\sigma_{z|B}^2$, (j) $\sigma_{z,w}$, (k) $\sigma_{z,w|B}$.

66. Random variables x and y have joint PMF shown in Fig. 4.20. Event $A = \{x > 0, y > 0\}$ and event $B = \{x + y \leq 3\}$. Determine: (a) p_x; (b) p_y; (c) $p_{x|y}(\alpha \,|\, 2)$; (d) $p_{y|x}(\beta \,|\, 4)$; (e) $p_{x,y|A^c \cap B}(\alpha, \beta \,|\, A^c \cap B)$; (f) $p_{x|A^c \cap B}$; (g) $p_{y|A^c \cap B}$; (h) whether or not x and y are independent; (i) whether or not x and y are independent, given $A^c \cap B$.

67. Random variables x and y have joint PMF shown in Fig. 4.20. Event $A = \{x > 0, y > 0\}$ and event $B = \{x + y \leq 3\}$. Determine: (a) $E(x)$, (b) $E(x^2)$, (c) σ_x^2, (d) $E(x - 2y)$, (e) σ_{2x}^2, (f) $E(5x - 3x^2)$, (g) $E(x \,|\, A \cap B)$, (h) $E(x^2 \,|\, A \cap B)$, (i) $E(3x^2 - 2x \,|\, A \cap B)$.

68. Random variables x and y have joint PMF shown in Fig. 4.20. Event $A = \{x > 0, y > 0\}$ and event $B = \{x + y \leq 3\}$. Determine: (a) $E(y)$, (b) $E(y^2)$, (c) σ_y^2, (d) $E(5y - 2x^2)$, (e) σ_{3y}^2, (f) $E(5y - 3y^2)$, (g) $E(x + y \,|\, A \cap B)$, (h) $E(x^2 + y^2 \,|\, A \cap B)$, (i) $E(3y^2 - 2y \,|\, A \cap B)$.

69. Random variables x and y have joint PMF shown in Fig. 4.20. Event $A = \{x > 0, y > 0\}$. If $w(x, y) = y - x$, then determine: (a) p_w, (b) $p_{w|A^c}$, (c) $E(w)$, (d) $E(w \,|\, A^c)$, (e) σ_w^2, (f) $\sigma_{w|A^c}^2$.

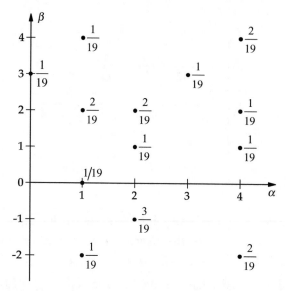

FIGURE 4.20: PMF for Problems 66–71.

70. Random variables x and y have joint PMF shown in Fig. 4.20. Event $A = \{x > 0, y > 0\}$. If $z(x, y) = xy$, then determine: (a) p_z, (b) $p_{z|A}$, (c) $E(z)$, (d) $E(z|A^c)$, (e) σ_z^2, (f) $\sigma_{z|A}^2$.

71. Random variables x and y have joint PMF shown in Fig. 4.20. Event $B = \{z + w \leq 1\}$, where $w(x, y) = y - x$, and $z(x, y) = xy$. Determine: (a) $p_{z,w}$, (b) p_z, (c) p_w, (d) $p_{z|w}(\gamma \,|\, 0)$, (e) $p_{z|B}$, (f) η_z, (g) $\eta_{z|B}$, (h) σ_z^2, (i) $\sigma_{z|B}^2$, (j) $\sigma_{z,w}$, (k) $\sigma_{z,w|B}$.

72. Random variables x and y have joint PMF shown in Fig. 4.21. Event $A = \{2 \leq x + y < 5\}$. Determine: (a) p_x, (b) p_y, (c) $p_{x,y|A}$, (d) $p_{x|A}$, (e) $p_{y|A}$.

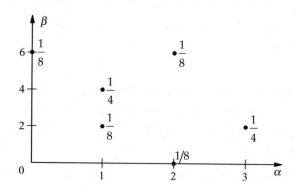

FIGURE 4.21: PMF for Problems 72–77.

73. Random variables x and y have joint PMF shown in Fig. 4.21. Event $A = \{2 \leq x + y < 5\}$. Determine: (a) $E(x)$, (b) $E(x^2)$, (c) σ_x^2, (d) $E(3x + 4x^2 - 5)$, (e) σ_{2x+5}^2, (f) $E(x \,|\, A)$, (g) $E(x^2 \,|\, A)$, (h) $E(3x + 4x^2 - 5 \,|\, A)$.

74. Random variables x and y have joint PMF shown in Fig. 4.21. Event $A = \{2 \leq x + y < 5\}$. Determine: (a) $E(3x + y)$, (b) $E(y^2 + x^2)$, (c) $E(4y + 3y^2 - 1)$, (d) σ_y^2, (e) $\sigma_{x,y}$, (f) σ_{3y+2x}^2, (g) $E(x + y \,|\, A)$, (h) $E(y \,|\, x = 2)$, (i) $E(x^2 + y^2 \,|\, A)$, (j) $\sigma_{y|A}^2$, (k) $\sigma_{x,y|A}$, (l) $\sigma_{x+y|A}^2$.

75. Random variables x and y have joint PMF shown in Fig. 4.21. Event $A = \{2 \leq x + y < 5\}$. If $w(x, y) = \max(x, y)$, then determine: (a) p_w, (b) $p_{w|A}$, (c) $E(w)$, (d) $E(w \,|\, A)$, (e) σ_w^2, (f) $\sigma_{w|A}^2$.

76. Random variables x and y have joint PMF shown in Fig. 4.21. Event $A = \{2 \leq x + y < 5\}$. If $z(x, y) = \min(x, y)$, then determine: (a) p_z, (b) $p_{z|A}$, (c) $E(z)$, (d) $E(z \,|\, A)$, (e) σ_z^2, (f) $\sigma_{z|A}^2$.

77. Random variables x and y have joint PMF shown in Fig. 4.21. Event $B = \{z - 2w > 1\}$, where $w(x, y) = \max(x, y)$, and $z(x, y) = \min(x, y)$. Determine: (a) $p_{z,w}$, (b) p_z, (c) p_w, (d) $p_{z|w}(\gamma \,|\, 0)$, (e) $p_{z|B}$, (f) η_z, (g) $\eta_{z|B}$, (h) σ_z^2, (i) $\sigma_{z|B}^2$, (j) $\sigma_{z,w}$, (k) $\sigma_{z,w|B}$.

78. Random variables x and y have the joint PMF shown in Fig. 4.22. Event $A = \{x < 4\}$, event $B = \{x + y \leq 4\}$, and event $C = \{xy < 4\}$. (a) Are x and y independent RVs? Are x and y conditionally independent, given: (b) A, (c) B, (d) C, (e) B^c?

79. Prove that if

$$g(x, y) = a_1 g_1(x, y) + a_2 g_2(x, y)$$

then

$$E(g(x, y)) = a_1 E(g_1(x, y)) + a_2 E(g_2(x, y)).$$

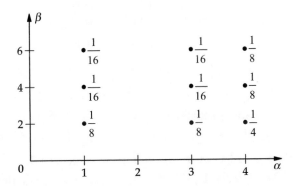

FIGURE 4.22: PMF for Problems 78.

80. Prove that if $z = g(x)$ then

$$E(z) = \sum_{\alpha} g(\alpha) p_x(\alpha).$$

81. Let event $A = \{g(x, y)\}$, where $g(x, y)$ is an arbitrary (measurable) function of the discrete RVs x and y. Prove or give a counter example:

$$p_{x|A}(\alpha \mid A) = \frac{p_x(\alpha)}{P(A)}.$$

82. Let event $A = \{g(x, y)\}$, where $g(x, y)$ is an arbitrary (measurable) function of the discrete RVs x and y. The RVs x and y are conditionally independent, given event A. Prove or give a counter example:

$$p_{x|A}(\alpha \mid A) = \frac{p_x(\alpha)}{P(A)}.$$

83. Random variables x and y are independent. Prove or give a counter example:

$$E\left(\frac{x}{y}\right) = \frac{E(x)}{E(y)}.$$

84. Random variables x and y are independent with marginal PMFs

$$p_x(\alpha) = \begin{cases} 1/3, & \alpha = -1 \\ 4/9, & \alpha = 0 \\ 2/9, & \alpha = 1 \\ 0, & \text{otherwise,} \end{cases}$$

and

$$p_y(\beta) = \begin{cases} 1/4, & \beta = 0 \\ 1/4, & \beta = 1 \\ 1/2, & \beta = 2 \\ 0, & \text{otherwise.} \end{cases}$$

Event $A = \{\min(x, y) \leq 0\}$. Determine: (a) $p_{x,y}$; (b) whether or not x and y are independent, given A; (c) $E(x + y)$; (d) $E(x + y \mid A)$; (e) $E(xy)$; (f) $\rho_{x,y}$; (g) $\rho_{x,y|A}$.

85. Random variables x and y satisfy: $E(x) = 10$, $\sigma_x = 2$, $E(y) = 20$, $\sigma_y = 3$, and $\sigma_{x,y} = -2$. With $z = z(x, y) = x + y$, determine: (a) $\rho_{x,y}$, (b) σ_{2x}, (c) $E(z)$, and (d) σ_z.

86. Random variables x and y satisfy: $\eta_x = 5$, $\eta_y = 4$, $\sigma_{x,y} = 0$, $\sigma_x = 4$, and $\sigma_y = 5$. Determine: (a) $E(3x^2 + 5x + 1)$, (b) $E(xy)$, (c) σ_{3x+2y}, (d) whether or not x and y are independent RVs.

87. A course in random processes is taught at Fargo Polytechnic Institute (FPI). Due to scheduling difficulties, on any particular day, the course could be taught in any of the rooms A, B, or C. The following *a priori* probabilities are known

$$P(A) = \frac{1}{2}, \quad P(B) = \frac{1}{3}, \quad P(C) = \frac{1}{6},$$

where events A, B, and C denote the events that the course is taught in room A, B, and C, respectively. Room A contains 60 seats, room B contains 45 seats, and room C contains 30 seats. Sometimes there are not enough seats because 50 students are registered for the course; however, they do not all attend every class. In fact, the probability that exactly n students will attend any particular day is the same for all possible $n \in \{0, 1, \ldots, 50\}$. (a) What is the expected number of students that will attend class on any particular day? (b) What is the expected number of available seats in the class on any particular day? (c) What is the probability that exactly 25 seats in the class will not be occupied on any particular day? (d) What is the probability that there will not be enough seats available for the students who attend on any particular day?

Besides having trouble with scheduling, FPI is also plagued with heating problems. The temperature t in any room is a random variable which takes on integer values (in degrees Fahrenheit). In each room, the PMF $p_t(\tau)$ for t is constant over the following ranges:

Room A: $70 \leq \tau \leq 80$,

Room B: $60 \leq \tau \leq 90$,

Room C: $50 \leq \tau \leq 80$;

outside these ranges, the PMF for t is zero.

(e) What is the PMF for the temperature experienced by the students in class? (f) Given that the temperature in class today was less than 75 degrees, what is the probability that today's class was taught in room A?

88. Random variables x_1 and x_2 are independent, identically distributed with PMF

$$p_{x_1}(\alpha) = \begin{cases} a/\alpha^2, & \alpha = -3, -2, 1, 4, \\ 0, & \text{otherwise.} \end{cases}$$

Random variable $y = x_1 + x_2$ and event $A = \{x_1 + x_2\}$. Find: (a) a, (b) $P(x_1 > x_2)$, (c) p_y, (d) $E(y)$, (e) $E(y \mid A)$, (f) σ_y^2, (g) $\sigma_{y \mid A}^2$.

89. Random variables x_1 and x_2 are independent, identically distributed with PMF

$$p_{x_1}(k) = \begin{cases} \dbinom{3}{k}(0.3)^k(0.7)^{3-k}, & k = 0, 1, 2, 3 \\ 0, & \text{otherwise.} \end{cases}$$

Find: (a) $E(x_1)$, (b) $\sigma_{x_1}^2$, (c) $E(x_1|x_1 > 0)$, (d) $\sigma_{x_1|x_1>0}^2$, (e) $P(x_1 \le x_2 + 1)$.

90. The Electrical Engineering Department at Fargo Polytechnic Institute has an out-standing bowling team led by Professor S. Rensselear. Because of her advanced age, the number of games she bowls each week is a random variable with PMF

$$p_x(\alpha) = \begin{cases} a - \dfrac{\alpha}{12}, & \alpha = 0, 1, 2 \\ 0, & \text{otherwise.} \end{cases}$$

To her credit, Ms. Rensselear always attends each match to at least cheer for the team when she is not bowling. Let $x_1, \ldots, x_n$ be n independent, identically distributed random variables with x_i denoting the number of games bowled in week i by Prof. Rensselear. Define the RVs $z = \max(x_1, x_2)$ and $w = \min(x_1, x_2)$. Determine: (a) a, (b) $P(x_1 > x_2)$, (c) $P(x_1 + x_2 + \cdots + x_x \le 1)$, (d) $p_{z,w}$, (e) $E(z)$, (f) $E(w)$, (g) $\sigma_{z,w}$.

91. Professor S. Rensselear, a very popular teacher in the Electrical Engineering Depart-ment at Fargo Polytechnic Institute, gets sick rather often. For any week, the probability she will miss exactly α days of days of lecture is given by

$$p_x(\alpha) = \begin{cases} 1/8, & \alpha = 0 \\ 1/2, & \alpha = 1 \\ 1/4, & \alpha = 2 \\ 1/8, & \alpha = 3 \\ 0, & \text{otherwise.} \end{cases}$$

The more days she misses, the less time she has to give quizzes. Given that she was sick α days this week, the conditional PMF describing the number of quizzes given is

$$p_{y|x}(\beta \,|\, \alpha) = \begin{cases} \dfrac{1}{4 - \alpha}, & 1 \le \beta \le 4 - \alpha \\ 0, & \text{otherwise.} \end{cases}$$

Let $y_1, y_2, \cdots, y_n$ denote n independent, identically distributed RVs, each distributed as y. Additionally, the number of hours she works each week teaching a course on probability theory is $w = 10 - 2x + y$, and conducting research is $z = 20 - x^2 + y$. Determine: (a) p_y, (b) $p_{x,y}$, (c) $p_{x|y}(\alpha\,|\,2)$, (d) $P(y_1 > y_2)$, (e) $P(y_1 + y_2 + \cdots + y_n > $

n), (f) $p_{z,w}$, (g) p_z, (h) p_w, (i) $p_{z,w|z>2w}$, (j) $E(z)$, (k) $E(w)$, (l) $E(z|z>2w)$, (m) σ_z^2, (n) $\sigma_{z|z>2w}^2$, (o) $\sigma_{z,w}$, (p) $\sigma_{z,w|z>2w}$, (q) $\rho_{z,w}$, (r) $\rho_{z,w|z>2w}$.

92. Professor Rensselaer has been known to make an occasional blunder during a lecture. The probability that any one student recognizes the blunder and brings it to the attention of the class is 0.13. Assume that the behavior of each student is independent of the behavior of the other students. Determine the minimum number of students in the class to insure the probability that a blunder is corrected is at least 0.98.

93. Consider Problem 92. Suppose there are four students in the class. Determine the probability that (a) exactly two students recognize a blunder; (b) exactly one student recognizes each of three blunders; (c) the same student recognizes each of three blunders; (d) two students recognize the first blunder, one student recognizes the second blunder, and no students recognize the third blunder.

Bibliography

[1] M. Abramowitz and I. A. Stegun, editors. *Handbook of Mathematical Functions*. Dover, New York, 1964.

[2] E. Ackerman and L. C. Gatewood. *Mathematical Models in the Health Sciences: A Computer-Aided Approach*. University of Minnesota Press, Minneapolis, MN, 1979.

[3] E. Allman and J. Rhodes. *Mathematical Models in Biology*. Cambridge University Press, Cambridge, UK, 2004.

[4] C. W. Burrill. *Measure, Integration, and Probability*. McGraw-Hill, New York, 1972.

[5] K. L. Chung. *A Course in Probability*. Academic Press, New York, 1974.

[6] G. R. Cooper and C. D. McGillem. *Probabilistic Methods of Signal and System Analysis*. Holt, Rinehart and Winston, New York, second edition, 1986.

[7] W. B. Davenport, Jr. and W. L. Root. *An Introduction to the Theory of Random Signals and Noise*. McGraw-Hill, New York, 1958.

[8] J. L. Doob. *Stochastic Processes*. John Wiley and Sons, New York, 1953.

[9] A. W. Drake. *Fundamentals of Applied Probability Theory*. McGraw-Hill, New York, 1967.

[10] J. D. Enderle, S. M. Blanchard, and J. D. Bronzino. *Introduction to Biomedical Engineering (Second Edition)*, Elsevier, Amsterdam, 2005, 1118 pp.

[11] W. Feller. *An Introduction to Probability Theory and its Applications*. John Wiley and Sons, New York, third edition, 1968.

[12] B. V. Gnedenko and A. N. Kolmogorov. *Limit Distributions for Sums of Independent Random Variables*. Addison-Wesley, Reading, MA, 1968.

[13] R. M. Gray and L. D. Davisson. *RANDOM PROCESSES: A Mathematical Approach for Engineers*. Prentice-Hall, Englewood Cliffs, New Jersey, 1986.

[14] C. W. Helstrom. *Probability and Stochastic Processes for Engineers*. Macmillan, New York, second edition, 1991.

[15] R. C. Hoppensteadt and C. S. Peskin. *Mathematics in Medicine and the Life Sciences*. Springer-Verlag, New York, 1992.

[16] J. Keener and J. Sneyd. *Mathematical Physiology*. Springer, New York, 1998.

[17] P. S. Maybeck. *Stochastic Models, Estimation, and Control, volume 1*. Academic Press, New York, 1979.

[18] P. S. Maybeck. *Stochastic Models, Estimation, and Control, volume 2*. Academic Press, New York, 1982.

[19] J. L. Melsa and D. L. Cohn. *Decision and Estimation Theory*. McGraw-Hill, New York, 1978.

[20] K. S. Miller. *COMPLEX STOCHASTIC PROCESSES: An Introduction to Theory and Application*. Addison-Wesley, Reading, MA, 1974.

[21] L. Pachter and B. Sturmfels, editors. *Algebraic Statistics for Computational Biology*. Cambridge University Press, 2005.

[22] A. Papoulis. *Probability, Random Variables, and Stochastic Processes*. McGraw-Hill, New York, second edition, 1984.

[23] P. Z. Peebles, Jr. *Probability, Random Variables, and Random Signal Principles*. McGraw-Hill, New York, second edition, 1987.

[24] Yu. A. Rozanov. *Stationary Random Processes*. Holden-Day, San Francisco, 1967.

[25] K. S. Shanmugan and A. M. Breipohl. *RANDOM SIGNALS: Detection, Estimation and Data Analysis*. John Wiley and Sons, New York, 1988.

[26] H. Stark and J. W. Woods. *Probability, Random Processes, and Estimation Theory for Engineers*. Prentice-Hall, Englewood Cliffs, NJ, 1986.

[27] G. van Belle, L. D. Fisher, P. J. Heagerty, and T. Lumley. *Biostatistics: A Methodology for the Health Sciences*. John Wiley and Sons, NJ, 1004.

[28] H. L. Van Trees. *Detection, Estimation, and Modulation Theory*. John Wiley and Sons, New York, 1968.

[29] L. A. Wainstein and V. D. Zubakov. *Extraction of Signals from Noise*. Dover, New York, 1962.

[30] E. Wong. *Stochastic Processes in Information and Dynamical Systems*. McGraw-Hill, New York, 1971.

[31] M. Yaglom. *An Introduction to the Theory of Stationary Random Functions*. Prentice-Hall, Englewood Cliffs, NJ, 1962.

Printed in the United States
by Baker & Taylor Publisher Services